Energy, Climate and the Environment

This is a critical time for nature-society relations as the global community faces the challenge of making food, energy, and transport systems more sustainable and resilient. Energy, climate and environmental systems are central to staying within planetary boundaries amid rapidly changing social, economic, political, and environmental conditions. This book series will focus on these topics, their interlocking impacts, interactions and uncertain outcomes. Contributions are encouraged from emerging and established scholars working within and/or across the energy, climate and environmental domains. The series provides an outlet for scholarship, debate, discussion, and dialogue between social sciences, humanities, physical and applied sciences. Its objective is to promote interdisciplinary and transdisciplinary research, whilst simultaneously fostering positive social change. The series provides an opportunity for scholars to connect state of the art science with the policymaking and practitioner community, accelerating the transition to a resilient future.

Connor McGookin · Evan Boyle

Engaging Communities in Climate Action

A Transdisciplinary Approach

palgrave
macmillan

Connor McGookin
University of British Columbia
Vancouver, BC, Canada

Evan Boyle
University College Cork
Cork, Ireland

ISSN 2947-8561 ISSN 2947-857X (electronic)
Energy, Climate and the Environment
ISBN 978-3-031-93497-1 ISBN 978-3-031-93498-8 (eBook)
https://doi.org/10.1007/978-3-031-93498-8

This Palgrave Macmillan imprint is published by the registered company Springer Nature Switzerland AG
The registered company address is: Gewerbestrasse 11, 6330 Cham, Switzerland

Preface

The mounting scientific evidence produced by the Intergovernmental Panel on Climate Change (IPPC) has yet to result in policy and actions that reflect the urgency of this crisis. At a global level, the 2015 Paris Agreement signalled a commitment to keep temperatures well below 2°C. However, efforts to outline pathways to the target have been inadequate to date. In response, there has been increasing calls for new scientific roles and approaches, to develop 'science-based' policy at various different levels, and also actively support the implementation of measures.

It is widely accepted that partnership and collaboration between both university and civic, community and private groups is crucial to the goal of addressing contemporary (un)sustainability challenges, and in particular climate change. This stems from two core goals; firstly, to understand the range of differing perspectives and contexts within which sustainability transitions are taking place, and secondly, the need for research to be societally relevant and support real-world solutions given the urgency of such threats.

Dingle Peninsula 2030 is an internationally recognised, awarding winning example of such a transdisciplinary collaboration for sustainability. As a flagship engaged research project within climate and energy fields, it provides many useful lessons for researchers and students seeking to build transdisciplinary partnerships for sustainability and climate action. This book follows the PhD journey of two early-career researchers from Energy Engineering and Sociology, working together with community partners, national bodies, and other groups on a regional development project. It outlines the institutional and practical challenges of doing such work; the story, successes, and trials of *Dingle Peninsula 2030*. It then closes with some advice, guidance, and reflections for other students/researchers setting out on a similar journey.

Vancouver, Canada Connor McGookin
Cork, Ireland Evan Boyle

Acknowledgements The authors are grateful to their colleagues in the MaREI Centre and University College Cork who supported the work: Prof. Brian Ó Gallachóir, whose leadership made it possible; Dr. Clare Watson for coordinating the many trips and meetings in Dingle; Aoife Deane for guidance and support; Prof. Edmond Byrne and Dr. Ger Mullay for supervision, and perhaps more important, laying the foundation in University College Cork for such collaborations to take place.

The book is informed by the *Dingle Peninsula 2030* project that we were part of during the course of our PhDs. Special thanks to our partners: Deirdre de Bhailís and Brendan Tuohy (Dingle Hub), Seamus O'Hara (NEWKD) and Claire McElligot (ESB Networks) and the wide network of local people on the peninsula who supported us and contributed their time and energy throughout our work. Additional thanks to Deirdre de Bhailís for the reflections that have been included in Chapter 10.

Competing Interests The authors have no competing interests to declare that are relevant to the content of this manuscript.

Ethics Approval Informed consent (to participate and to publish) was obtained from the individual interviewed as part of Chapter 10.

Contents

About the Authors

Connor McGookin is a Postdoctoral Research Fellow with the Systems Beings Lab at the University of British Columbia, Canada. His central research focuses on ways to improve climate and energy planning processes by engaging key groups such as local governments and communities.

He completed his doctoral thesis "*Developing participatory methods in energy system modelling and planning*", while being a member of the award-winning Dingle Peninsula 2030 and Imagining 2050 projects. His main research interest was to explore ways of working across social/technical disciplines and with a diverse range of groups to collaboratively form energy and climate scenarios. Connor has since sought to support local government climate action planning with several projects both in Ireland and British Columbia, Canada. Current research interests include assessing spatial elements of the energy system with GIS tools and the integration of energy justice principles in energy planning. A recognised leader in the field of participatory energy systems modelling, he sits on the scientific advisory committee for the International Renewable Energy Agency (IRENA)'s Participatory Processes for Long-term Energy Scenario Development.

Evan Boyle is a Senior Postdoctoral researcher with the Sustainability Institute, University College Cork. His research focuses on community engagement on energy transitions, transdisciplinary approaches to sustainability, and city-university partnerships for climate action.

Previously, he has investigated public participation in renewable infrastructure development in both Ireland and internationally through the *STEPS project* (2022–2024) and just transition dimensions of city level climate action through *JustCities Hub* (2024–2028). As of 2025, he acts as principal investigator on *UrbanClimateCollab*, exploring city-university collaborations for climate action (2025–2029). Other current research interests include transdisciplinary methodologies, blue spaces and nature immersion, and liminality and ritual festivals. He has recently published work in *Theory, Culture and Society, Energy Research & Social Science,* and *Energy Policy.* He is a member of the Sociological Association of Ireland and a working group leader on the COST Action SHiFT. He is co-founder of the *Mission Restore our Ocean and Waters Charter* action 'Blue Connections', a multidisciplinary conference and consortium exploring societies connection to the ocean. Evan is on the Editorial Board for the journal *Societal Impacts.* He is a commissioner on Ireland's Just Transition Commission with secretariate functions provided through the Department of Climate, Energy and the Environment, Government of Ireland.

Abbreviations

AD	Anaerobic Digestion
BEI	Baseline Emission Inventory
CAP	Climate Action Plan
CHP	Combined Heat & Power
COP	Conference Of the Parties
CSO	Central Statistics Office
DZ	Decarbonisation Zone
EMP	Energy Masterplan
ESRI	Economic & Social Research Institute
EU	European Union
EV	Electric Vehicle
FFD	Fossil Fuel Divestment
GAA	Gaelic Athletic Association
GHG	Greenhouse Gas
ICCA	Ireland's Climate Change Assessment
IP	Interview Participants
IPCC	Intergovernmental Panel on Climate Change
ISIF	Irish Strategic Investment Fund
JOCCA	Joint Oireachtas Committee on Climate Action
LEAP	Low Emissions Analysis Platform

LED	Light Emitting Diode
LPG	Liquid Petroleum Gas
MaREI	The SFI research Centre for Energy, Climate and Marine
NDCs	Nationally Determined Contributions
NEWKD	North East West Kerry Development
NGO	Non-Governmental Organisation
PV	Photovoltaic
RESS	Renewable Electricity Support Scheme
SDGs	Sustainable Development Goals
SEAI	Sustainable Energy Authority of Ireland
SEC	Sustainable Energy Community
SECAP	Sustainable Energy & Climate Action Plan
SFI	Science Foundation Ireland
TdR	Transdisciplinary Research
UCC	University College Cork
UN	United Nations
UNFCCC	United Nations Framework Convention on Climate Change

List of Figures

List of Tables

1

Introduction

Connor McGookin and Evan Boyle

1.1 Opening Remarks

Embarking on the journey of a PhD can come with a fair amount of apprehension regarding the challenges that lie ahead. These concerns can intensify when attempting to collaborate with partners outside of academia. Beyond the familiar walls of the university, one faces numerous hurdles. Fortunately, by coincidence, we were both pursuing our PhDs at the same time, are from the same county in Ireland, and often spent time together in Cork City (playing Catan). Strangely, we didn't realise we were both starting on this journey together until the first day we met in the office. This happy coincidence greatly facilitated the development of our transdisciplinary, action-oriented research project.

During our time as PhD students, despite high-level institutional support for such practices, there was a lack of training opportunities available. There was a lack of suitable modules for us to undertake within the university that would give us the grounding we needed in inter/transdisciplinary practice. Instead, we learnt how to collaborate by getting together and figuring it out as the project developed, while practical facilitation skills were gained through professional training courses

C. McGookin and E. Boyle, *Engaging Communities in Climate Action*, Energy, Climate and the Environment, https://doi.org/10.1007/978-3-031-93498-8_1

aimed at community development workers. The disconnect between institutional mantras highlighting support for such approaches and the lack of any practical support or guidance caused us much frustration as we were getting to grips with the project. We've since been keen to support other interested early-career researchers by developing supports and resources. The primary purpose of this book is to summarise our work as part of *Dingle Peninsula 2030*, and provide guidance for other prospective students looking to pursue similar inter/trans-disciplinary collaborations.

There are now some notable examples of transdisciplinary programmes available such as the TdLab in ETH Zurich, the PhD programme in Transdisciplinary Sustainability hosted by the Memorial University of Newfoundland, or the Transdisciplinary Collaborative PhD cohort run at the University of British Columbia. There are also a growing number of toolkits [1, 2], guides, and handbooks [3–6] now available to researchers seeking to pursue transdisciplinary approaches. Our hope is to contribute to this budding area of scholarship by documenting our experience as part of a transdisciplinary research project. The book follows our PhD journeys, critically reflecting on the transdisciplinary approach taken, as well as outlining a very practical guide for how to engage communities in climate action and sustainability issues.

1.2 Introducing the Book

We begin the book with an overview of transdisciplinary research based on our collective literature reviews as part of our doctoral theses and interviews with more senior colleagues undertaking similar work. For over a decade now, transdisciplinary research has cemented itself as a means through which the complex nature of societal grand challenges can be analysed, understood, and acted upon in a more holistic manner. In light of pressing (un)sustainability challenges like climate change and biodiversity loss, interest in transdisciplinary research has grown as the need for real-world impact has become increasingly important. Chapter 2 discusses the historical roots of transdisciplinary research (TdR), its growing prominence in funding priorities and university strategies, as well as some of the challenges associated with working both across disciplines and with different partners. Understanding the institutional context within which we must operate and pitfalls faced by transdisciplinary researchers provides a useful foundation upon which the rest of the book is built. The chapter offers an introduction to transdisciplinary research for students setting out on a similar path, providing an extensive collection of relevant literature on the topic as well as critical reflections from researchers at various career stages.

Having outlined how our approach is grounded within the literature on transdisciplinary research, we then summarise what makes a suitable case study area for our investigation. The question at the heart of our PhD investigations was what are the societal barriers to the decarbonisation of our energy system. There has been great interest in the potential for cities or regions to pursue energy independence or autarky, and demonstrate leadership on decarbonisation efforts. There are a range of different initiatives throughout Europe, for example: 'energy self-sufficient regions' in the European Alps, '2000-W society' in Switzerland, 'Transition Towns' in the United Kingdom 'bioenergy villages' in Germany, 'energy regions' in Austria, and 'renewable energy islands' in Denmark. The shared goal between all these initiatives is to reduce dependence on fossil fuel imports in place of local renewable sources. Asides from the obvious environmental benefits such a transformation brings, most important is the socio-economic benefits. The key motivator behind such initiatives is not so much the desire to reduce their carbon footprint, but rather to rejuvenate the local economy, which is particularly important in struggling rural areas. Chapter 3 will look at four exemplar community efforts to reduce fossil fuel dependence: Güssing, Austria, Samsø Island, Denmark, Jündhe, Germany, and Orkney Island, Scotland. While the manner in which each community went about transforming their energy system differs in terms of key actors or technologies pursued, there are several common themes and key learnings apparent. The chapter serves as a good practice guide for researchers and other interested people seeking to develop community energy or regional energy transition projects.

Before getting into the focus of this book, *Dingle Peninsula 2030*, Chapter 4 details the wider national context within which the project takes place. It involves a genealogical outline of environmental narratives in the Irish context from 2006 to 2024. The purpose of discussing narratives is to explore the storylines which are developed through the economy, in media, policy and socio-political debates, framing the context in which environmental issues are understood in Ireland. Environmental narratives throughout the twenty-first century have helped to shape the reality of the national interpretation of environmental issues. These narratives, however, have not remained static. Shifts in framing on environmental issues and sustainability have been borne out of numerous factors, be they internal or external, civil or political. Our focus on the years 2006–2024 covers the publication of the Stern report and the last days of the Celtic Tiger, through the recession, up to the Paris Accord, and beyond to Fridays for Future, Extinction Rebellion, COVID-19 and the War on Ukraine. The chapter is prefaced with a brief outline of the previous three decades, back to the 1970s. Through an analysis of policy, civil society developments and media messaging, twenty-first-century environmental narratives in Ireland are outlined, giving a national context to the specific investigation provided here.

Chapter 5 then provides a summary of the Dingle Peninsula historically, culturally, and socially. Definitions of community have long been contested from the earliest sociological accounts provided by Tönnies in *Gemeinschaft and Gesellschaft* and the work of Weber, to theorists of the Chicago School such as Park and Wirth. Despite the long-standing debate on community within sociology, within this chapter an outline is provided which helps to refine the application of the term to the *Dingle Peninsula 2030* project. The project is borne out of a notion of community as desire. The desire relates to rectifying some of the issues of unsustainability found within the region. The chapter reflects upon the project as a collaborative endevour aiming to transition the region to a low-carbon, sustainable community by 2030 and how this relates to questions of 'community'.

A vital first step for regional energy transitions is to develop an understanding of the current energy balance and related carbon dioxide emissions in order to develop a decarbonisation plan. Chapter 6 covers the experience of co-producing energy strategies on the Dingle Peninsula. It includes discussion of both the initial baselining exercise undertaken to establish current energy-related CO_2 emissions and the scenarios developed to explore different pathways to 2030. The baseline clearly highlights the challenge decarbonisation poses for rural areas with a heavy reliance on imported oil, and a dispersed population driving higher heating and private transport demand. This analysis informed the projects pursued by our local partners with transport seen as a critical area to address. Building on the baseline, scenarios were then modelled using a simulation tool (Low Emissions Analysis Platform, LEAP) to explore how the energy system might change over the next decade. This was informed by extensive community engagement. Attention is given to how the community inputs not only shaped the analysis conducted but also led to a complete revision of the research plan. The initial approach for linking community inputs and the modelling tools based on existing examples in the literature was scrapped in favour of a broader discussion of the social, economic, and environmental challenges the area faced. This posed a serious methodological challenge but was essential to align with the interests of our local partners and respect the issues facing the community. The chapter offers reflections on how well models can represent community interests, the insights gained from the exercise, and learnings for others seeking to develop energy strategies in collaboration with a diverse group.

Initially conceived as a project focused on the energy transition, *Dingle Peninsula 2030* has developed beyond the project's original scope, with a variety of wider sustainability-related projects, initiatives and actions emerging. Effective climate action necessitates a holistic approach. This flowering of wider action has been termed the diffusion of sustainability. The multi-stakeholder collaboration has acted as a springboard for local efforts, which stands to highlight the key role which transdisciplinary action-orientated research approaches can play in helping to support regional sustainability across a range of sectors. Beyond local impact, *Dingle Peninsula 2030* has received a number of prestigious awards both within the Academy and at a national level. It has shared insights with national policymakers, been invited to COP-26 in Glasgow and outlined by the UN Regional Information Centre as an exemplar project for regional sustainability. Due to the clear impact of the project, there has been a great interest from a range of different groups concerned with sustainability transitions. A common question that is often asked is: "*How can we do this elsewhere?*" Chapter 7 outlines the diffusion of sustainability and the concept of 'scaling deep' as a means through which the success of *Dingle Peninsula 2030* can be understood.

Chapter 8 is a collection of the various engagement activities that were carried out over the four years, grouped by an adaptation of the Wellcome Trust's 'Engagement Onion' framework. It groups activities into three categories based on purpose of the engagement; 1) communicating—disseminating findings from the project and raising awareness of the group's activities, 2) consulting—asking people about their thoughts on different issues, and 3) collaborating—involving people in the decision-making process or coordinating events with our local partners. The purpose of this chapter is to act as a useful guide for researchers and others interested in engaging the wider public in discussions about climate change and our future energy system. There were some novel events developed during the project such as the LED bulb swap and climate hacks. For each event or activity, a summary of how it was run is provided along with reflections on the areas for improvement.

In Chapter 9, we reflect on the experience of adapting to an inter/trans-disciplinary work environment. Picking up on some of the issues raised in Chapter 2, it explores the various barriers to pursuing such approaches, offers some critical reflections on the contribution the research made to the *Dingle Peninsula 2030* project, and how the role of research is changing. The main contribution of the chapter is to outline recommendations for future students seeking to pursue transdisciplinary approaches, and ways in which the institutions or funding bodies that wish to promote such approaches can provide support.

In Chapter 10, we outline some final remarks on the Dingle Peninsula as our study area, we discuss the role of research in responding to the societal challenges presented by climate change, and situate our investigation in relation to broader trends towards new means of responding to these problems. To close this chapter and the book, the final words are given to Deirdre de Bhailís, manager of Mol Téic/Dingle Hub to discuss how things have evolved since the core research project finished in 2022, and take a look towards the future of the peninsula.

1.3 Weaving a Strand Through the Book: *Dingle Peninsula 2030*

As will be explored in Chapter 3, regional energy transitions have tended to take place in isolated rural areas. The social and economic challenges these areas face is the backdrop that spurs a commitment to reduce spending on imported fossil fuels in favour of local renewables. In Ireland, similar challenges can be seen throughout rural areas within the Border, Midlands and Western regions. One particularly interesting region is the Dingle Peninsula, a small isolated 583 km^2 peninsula about 50 kilometres long and 15 kilometres wide, with a population of ~ 12,800. Despite being one of Ireland's most popular tourist destinations, the area is struggling to provide a home for young families and faces serious economic challenges (see Chapter 5, and in particular 5.2 Background). With regard to energy, the area is highly dependent on important fossil fuels (Chapter 6), and often experiences blackouts as there is just a single electricity line running down the peninsula.

Dingle Peninsula 2030 or *Corca Dhuibhne 2030* (Irish translation) was a collaboration established with the aim to set the region on the path to a low-carbon, sustainable society by 2030. Locally, the preferred name for the project was *Corca Dhuibhne 2030* in light of the regions status as a Gaeltacht, an area of Ireland where Gaeilge (Irish) is still the first language. However, throughout this book the English version, *Dingle Peninsula 2030* is used as it might be more widely recognised. The organisations involved were the not-for-profit, Dingle Creativity and Innovation Hub (Dingle Hub/Mol Téic); the local community development organisation, North East West Kerry Development (NEWKD); Ireland's national electricity distribution system operator, ESB Networks, and MaREI, the SFI research Centre for Energy, Climate and the Marine. The collaborative committee formed included seven members: a board member of Mol Téic, the manager of Mol Téic, the Local Area Manager within NEWKD, ESBN Network's Dingle Community Engagement Manager, a community engagement specialist within MaREI and two PhD students. Alongside this, the Communications and Public Engagement Manager within MaREI acted as an associate member of the committee. A transdisciplinary approach was taken throughout, working across University College Cork's Department of Sociology & Criminology and the School of Engineering, and within the community context of the Dingle Peninsula through the collaborative committee.

From a research perspective, the area was identified as providing an opportunity to explore an 'engaged research' approach due a couple of exciting developments taking place simultaneously. The national distribution system operator (ESB Networks) was running a pilot project (The Dingle Project) to explore the electrification of heating and transport, with local ambassadors trailing technologies such as air-sourced heat pumps, electric cars, battery storage systems, and solar PV. In line with this, a community-led group based out of the Dingle Creativity & Innovation Hub (Mol Téic), located within the town of Dingle, was beginning to investigate the broader societal changes required through a range of initiatives with the primary goal of understanding how the ongoing digital transformation and sustainability transition may offer the local community new and interesting employment opportunities, to diversify away from an over-reliance on tourism. NEWKD emerged as an important contributor due to their standing within the community and their area managers' involvement with other community groups like Transition Kerry. Through initial conversations in 2017, the potential for a collaborative approach to a project which would align our collective ambitions under a shared vision began to percolate before formalising the following year. From this point, things moved quickly, with *Dingle Peninsula 2030* established as a vehicle to explore the decarbonisation of the region and the range of interrelated opportunities and challenges that this represented both locally and nationally. Throughout the process, our research team employed a variety of different engagement activities (Chapter 8), grounded in an engaged research or action-orientated approach, as outlined below by the Irish University Association, Campus Engage:

> *Engaged research describes a wide range of rigorous research approaches and methodologies that share a common interest in collaborative engagement with the community and aim to improve, understand or investigate an issue of public interest or concern, including societal challenges.* [7, p. 28].

The development of such a collaboration enabled the emergence of a network of local, national, and international contacts which could be partnered with, strengthening each individual organisation in the partnership through taking such an approach. Alongside this, the different individual and organisational capacities and skills developed a strong holistic base of expertise for seeking to deliver different initiatives within the collaboration. Despite these clear benefits, challenges existed with regard to merging different institutional and organisational cultures. As such, trust-building was key. Our internal organisational structures, like most, were often poorly equipped to manage this need for time or trust-building. While the establishment and achievement of individual goals can be realised, the ability of a range of organisations to identity a shared vision was a challenge for our group, while also being a key consideration within the literature [8]. Trust-building has been linked previously to power sharing [9], to help manage power asymmetries within collaborative relationships [9, 10]. The need to create space, through freeing up time, to enable trust-building is of central importance to collaborative governance processes.

The desired outcomes or goals being pursued by each individual organisation can limit the scope for collaboration. There were several key differences among us as partners. ESB Networks, the electricity distribution system operator, are solely focused on electricity. However, the research of MaREI and initiatives emerging from Mol Téic seek to address the whole energy system, including alternatives to electrification in heating and transport. The local organisations (NEWKD/Mol Téic) have a long-term vision and commitment to the area, whereas for the public body and research institute (ESBN/MaREI) involvement is much shorter, perhaps only lasting the duration of the project. A change is needed with relation to the way in which institutions interact with community groups and citizens to effectively go about implementing the measures needed to expedite the transition to a zero-carbon energy system. Institutions have a key role to play within sustainable transitions and within this institutional change is considered of utmost importance [11–13]. We have acknowledged the benefits of our group, as a collaboration where the whole is greater than the sum of the parts. Yet, we still must make clear the difficulties around establishing such a group, and the need to create mechanisms within public bodies and research institutes to participate in such collaborations to deliver on sustainable transition ambitions. These issues are explored in detail in Chapter 9.

A key question faced by transdisciplinary research is what was the impact of the project? Evaluation of the research activities and how they support real-world action is a critical area for investigation. Our group has collectively developed initiatives across several sectors and as time went by a range of different sustainability projects emerged leading to our coining of the term 'the diffusion of sustainability' through the process for regional development of 'scaling deep'. As noted, these concepts are outlined in Chapter 7, which examines how the project developed and what initiatives emerged. The lasting impacts of the project are discussed in the concluding Chapter 10. Through conversation with the manager of the Dingle Hub, we look at changes in how they operate since the direct involvement of MaREI and ESBN ended, and what this means in terms of the ultimate goal to be a climate action and sustainability leader.

References

1. G. Bammer, Toolkits for transdisciplinarity (GAIA – Ecological Perspectives in Science and Society, 2017). https://www.oekom.de/_uploads_media/files/gaia_flyer_toolkits_032911.pdf
2. td-net, Network for transdisciplinarity research, methods and tools for co-producing knowledge (2020). https://naturalsciences.ch/co-producing-knowledge-explained. Accessed 14 Nov 2024
3. G.H. Hadorn, H. Hoffmann-Riem, S. Biber-Klemm, W. Grossenbacher-Mansuy, D. Joye, C. Pohl, U. Wiesmann, E. Zemp (eds.), *Handbook of Transdisciplinary Research* (Springer Netherlands, Dordrecht, 2008). https://doi.org/10.1007/978-1-4020-6699-3
4. M. Padmanabhan, *Transdisciplinary Research and Sustainability* (Routledge, London, 2017). https://api.taylorfrancis.com/content/books/mono/download?identifierName=doi&identifierValue=https://doi.org/10.4324/9781315441481&type=googlepdf. Accessed 14 Nov 2024
5. P. Leavy, *Essentials of Transdisciplinary Research: Using Problem-Centered Methodologies* (Routledge, 2016). https://www.taylorfrancis.com/books/

mono/https://doi.org/10.4324/9781315429137/essentials-transdisciplinary-research-patricia-leavy. Accessed 14 Nov 2024

6. R.J. Lawrence, *Handbook of Transdisciplinarity: Global Perspectives* (Edward Elgar Publishing, 2023). https://books.google.com/books?hl=en&lr=&id=2ki-EAAAQBAJ&oi=fnd&pg=PR1&dq=Handbook+of+Transdisciplinarity:+Global+Perspectives&ots=iQNmeDda12&sig=JHR2X0Jdgf9GSChsNMYlcdaR-Ms. Accessed 14 Nov 2024
7. Campus Engage IUA, *Engaged Research* (Irish Research Council, Dublin, 2017). https://research.ie/assets/uploads/2017/04/Engaged-Research.pdf
8. P.W. Mattessich, B.R. Monsey, *Collaboration: What Makes It Work.* A Review of Research Literature on Factors Influencing Successful Collaboration (ERIC, 1992)
9. B. Ran, H. Qi, Issues and challenges of public service procurement in China: a collaborative governance perspective. Int. J. Public Policy **12**, 339–358 (2016)
10. J.M. Purdy, A framework for assessing power in collaborative governance processes. Public Adm. Rev. **72**, 409–417 (2012)
11. F.W. Geels, From sectoral systems of innovation to socio-technical systems: insights about dynamics and change from sociology and institutional theory. Res. Policy **33**, 897–920 (2004)
12. S. Jolly, R.P.J.M. Raven, Collective institutional entrepreneurship and contestations in wind energy in India. Renew. Sustain. Energy Rev. **42**, 999–1011 (2015)
13. S. Wirth, J. Markard, B. Truffer, H. Rohracher, Informal institutions matter: professional culture and the development of biogas technology. Environ. Innov. Soc. Transit. **8**, 20–41 (2013)

2

Transdisciplinarity for Sustainability: Reflections on Experience to Guide Practice

Evan Boyle and Connor McGookin

2.1 A Question of Transdisciplinarity?

There is an increased belief among researchers, stakeholders and funding bodies that co-produced knowledge can lead to enhanced opportunities for meaningful progress across policy, decision-making and practice about sustainability issues [1]. As outlined by Funtowicz and Ravetz [2], given the high stakes and uncertainty presented by contemporary ecological challenges, a 'post-normal science' approach is required to incorporate experiential (e.g. community, indigenous, local) knowledge.

In parallel to this, and in large part emanating from it, there has been growing recognition of the value of transdisciplinary (TdR) approaches to research to respond effectively to sustainability-related issues [3] [4] [5] [6] [7] including across higher education [8], and also to ensure societal impact "*is most explicitly expressed in, and actively shaped by transdisciplinary research programmes*" [9]. Standard research funding, however, can present challenges whereby the landscape found within funding calls encourages utility over the critical interrogation of the dynamics at play.

C. McGookin and E. Boyle, *Engaging Communities in Climate Action*, Energy, Climate and the Environment, https://doi.org/10.1007/978-3-031-93498-8_2

At the turn of the century, a call emerged for joining up scientific and technological knowledge with political and social processes concerning complex interconnected systems related to sustainability [10, 11], and was followed by the emergent field of sustainability science. An importance was placed on understanding global processes and their relationship to localised social and ecological considerations. Traditional neo-Cartesian approaches to scientific investigation, specifically the application of the scientific method involving a 'detached' or 'neutral' researcher conducting their studies, were called into question by this new approach. New forms of learning and knowing were championed as providing mechanisms to enable multi-actor collaboration across different sectors or social groupings in addressing the complex challenges related to sustainability [11].

Stemming from the call for new forms of learning and knowing is the need for new forms of research "that generate knowledge that matters to people's decisions and engages in arenas where power dominates knowledge" ([12] P. 1). A plethora of research approaches have emerged or been re-imagined in light of this need for co-produced actionable knowledge, often under the guise of transdisciplinarity, influenced by the tradition of Mode 2 science. This has led to a call for an improvement in collecting and reporting on these processes as a measure of their efficacy and impact [13], and on from this active learning through reflective practice [14].

It is now widely accepted that partnership and collaboration with both academic and non-academic stakeholders are crucial to the goal of generating real-world impact, which lies at the heart of sustainability science [15, 16]. This has drawn into question the role of the researcher, with those pursuing a transdisciplinary approach taking on a range of new roles such as the reflective scientist, process facilitator, knowledge broker, change agent and self-reflexive scientist [17]. Partnerships with actors outside of academia greatly expand the expectations placed upon researchers compared with traditional roles; assuming responsibility for activities such as guiding collective learning activities [18] or implementing sustainability solutions in practice [19]. Since little attention has been given to the challenges of undertaking these roles [20], practising researchers are invariably left "*without the proper vocabulary to explain and navigate the tensions and potentials that come with their 'new' activities and roles*" ([21] p. 484). Thus, it is important to build a better understanding of researcher's experiences in these variety of roles in order to develop suitable training and development opportunities [22].

This builds upon the notion of 'the honest broker' [23], whereby the positionality of scientists in relation to policy formation is developed, with the honest broker developed in contrast to the science arbiter, the pure scientist, and the issue advocate. The honest broker, as such, is characteristically defined through their "*effort to expand (or at least clarify) the scope of choice for decision-making in a way that allows for the decision maker to reduce choice based on his or her own preferences and values*" ([23] p. 2–3). Building on this, the notion of honest brokering has been related to images of knowledge rather than knowledge itself addressing beliefs held on the nature of truth, the task of science, the task of academic and non-academic participants, and the task of transdisciplinary research [18]. It must be noted, however, that the honest broker role has been critiqued for restricting the potential for the reflexive scientist [24] (discussed further in Chapter 9).

Within the disciplinary context of sociology, debate has long since continued about the role of the discipline, and the different traditions contained within; what it sets out to achieve and for what reasons [25]. Some debate about the future of sociology has centred on two distinct but competing interpretations of the discipline's purpose and relevance: transdisciplinarity and tradition [26]. One argument in favour of a transdisciplinary approach is concerned with the complexity of contemporary phenomena and their multitude of layers and multifaceted realities. Siloed disciplinary boundaries then cannot comprehensively, or even sufficiently in some cases, provide an interpretation of the complexity of dynamics at play [27]. One argument in favour of the traditionalist approach states that a move towards transdisciplinarity leaves sociologists in a more vulnerable position, weakening the strength of the discipline [28]. The future of the discipline perhaps lies somewhere in the middle of these two positions whereby a transdisciplinary approach can be undertaken in certain contexts, while in others a more traditionalist approach is sought, with both modalities informing the other in an iterative process.

The concept of a new more socially responsible engineer who serves the community as oppose to an employer or client was originally proposed by Sharon Beder in her book 'The New Engineer' [29]. With an increasing role for engineers in public policy and planning, she outlines the need for greater reflection on the role of engineering in society, as well as engineering practice and ethics in dealing with risk and uncertainty. In her concluding remarks, she notes that: "*Such interaction with the community is necessary to improve the general understanding of the role of engineering and ensure that the engineering viewpoint is considered by the nation's decision-makers. It is also important that such comment not be disguised or presented as neutral technical advice, but as the value-based judgment it necessarily is*" [29]. More recently, Conlon (2008) highlights that engineers need to understand the wider social context in which they work, and reflect on how their involvement with public policy may "*enable or constrain the move towards a sustainable and just world*" [30]. Many of the barriers to climate mitigation are non-technical, and therefore, cannot be addressed through traditional reductionist engineering approaches [31]. Reflecting on how engineering can respond to super wicked problems such as climate change, Yearworth concludes that *"Dealing with worldviews, subjectivity and the fact that we are contributing to the problem in which we are trying to intervene suggests that we need to extend existing methods to deal with more inclusive and widespread participation"* [32].

Defining Transdisciplinarity

Transdisciplinarity, as with many academic concepts, has lent itself to several conceptions, including two conceptual "strands" [33], involving at least five discernible major "clusters of meaning" [34]. For many, a key aspect is the use of a *process* through which the production of knowledge is opened up to a wide variety of perspectives and viewpoints [35]. While Klein ([34] p. 189) posits that "*there is no universal theory, methodology, or definition of transdisciplinarity*", it is useful to ground ourselves somewhere within the discussion. Firstly, an understanding of the difference between multi-, inter- and transdisciplinary approaches. This has been developed previously within the work of Mullally et al. [33]. In a cursory manner, the discussion can be summarised as such; interdisciplinarity is set apart from multidisciplinarity as requiring deep engagement between a number of disciplines. This includes an understanding of respective terminologies and methods by the participants. It differs from transdisciplinarity which is commonly involving "*not only multiple disciplines but also multiple non-academic participants (e.g., land managers, user groups, the general public) in a manner that combines interdisciplinarity with participatory approaches*" [36, p. 9].

Additionally, a transdisciplinary approach is best suited to complex issues which cannot be easily classified as the domain of one perspective or discipline over another [37]. It is therefore seen as most appropriate in dealing with complexity and complex systems, as such an approach helps to overcome narrow specialisation, a feature which has risen to prominence within academia since the late nineteenth and early twentieth century [35]. As Balsiger [38] suggests, a transdisciplinary approach to research should move beyond any pre-established disciplinary frameworks, instead seeking to consolidate different forms of knowledge both inside and outside the academy to address the complex issue at hand (Table 2.1).

Transdisciplinary research encompasses a broad field of study and understanding. Thus, it is worth noting some guiding principles. The seminal work of Lang et al. (2012) provided one of the first outlines of transdisciplinary research principles from their review of a number of early examples [20]. It included:

Table 2.1 Overview of commonly stated reasons that a transdisciplinary approach is needed

Description	References
High stakes and uncertainty call for 'post-normal science' that includes experiential knowledge	[39]
Real-world knowledge is needed to understand the diversity of problem framings	[40]
A process of mutual learning between science and society to produce new knowledge	[41]
A means of integrating different types of knowledge for local environmental management	[42]
Research needs to generate usable knowledge that matters to people's decisions	[12]
Given the urgency of sustainability issues, research must contribute to real-world solutions	[43–45]

- Establishment of a transdisciplinary committee with a diverse group of academic and non-academic representativeness
- Jointly define the problem, research question(s) and process
- Generate targeted products for both parties
- Evaluate scientific and societal impact

More recently, Fazey (2018) provide ten essentials for action-orientated climate and energy research [45], as outlined in Table 2.2.

With climate change impacting on complex systems at local, regional, national, and global levels, and considering the urgency needed and inertia present (i.e. a super wicked problem), the merits of a transdisciplinary research approach are evident. While there are difficulties in conducting transdisciplinary research [46–48] it would seem evident that not only can it be implemented, but also that if done effectively it will lead to far more impactful, insightful, and progressive findings then what is currently being offered. However, despite the increased importance being placed on transdisciplinary approaches (in particular within funding calls), the management and evaluation practices are still underdeveloped [49]. The challenges associated with practising TdR and ways to support the emergence of the research approach have received plenty of attention (Table 2.3).

Table 2.2 Ten essentials for action-oriented and second-order energy transitions, transformations and climate change research (Fazey et al. 2018)

1) Focus on transformations towards low-carbon, resilient living
2) Focus on solution processes
3) Focus on 'how to' practical knowledge
4) Approach research as occurring from within
5) Work with normative aspects
6) Seek to transcend current thinking and approaches
7) Take a multifaceted approach to understand and shape change
8) Acknowledge the value of alternative roles of researchers
9) Encourage second-order experimentation and change
10) Be reflexive

Table 2.3 Summary of challenges, motivations and ways forward for practitioners of TdR within the literature

Challenges		Motivations	Ways forward
Different linguistic traditions [50]	Difficulty gaining influence in policy [51]	Advancement of common good [52]	New transdisciplinary research institutes [53]
Tension between technical expertise and organisational structures [50]	Career progression opportunities [54]	Desire to engage with important issues [52]	Spaces for conversation on transdisciplinarity [7]
Use of traditional evaluation mechanisms [55]	Funding for trust-building [13]	Sense of real-world impact [52]	Researcher and institutional leadership, as well as establishing a shared understanding of public engagement [56]
Specialised knowledge and disciplinary silos [57]	Failing to acknowledge existing stakeholder engagement practices [58]	To inspire learning, develop researchers' skills; be ethical, accountable and transparent; make the world a better place; create a more efficient, dynamic and sustainable economy; enhance social cohesion and democratic participation; increase the quality and impact of research, and; win support for science [59]	

(continued)

Table 2.3 (continued)

Challenges		Motivations	Ways forward
Time requirements [44]	Evolving role of researchers [18, 21]	Social responsibility and problem-solving [60]	
Diversity of organisational expectations [61]			

2.2 Questioning Transdisciplinarity

In setting out to question approaches to transdisciplinarity and the different experiences of such, we gathered together a cohort of academics within University College Cork who had relevant experience. The methodological approach taken and the in-depth analysis of findings has been published as a peer-reviewed article [62]. While nascent, there is an emergent importance placed on TdR within our University, as illustrated through two edited collections on the topic [7, 63]. As noted by Ni Shuilleabhain et al., [56] "*Universities in Ireland have not been immune to the international trends focusing on supporting, funding and embedding public engagement in university teaching, research and culture*". University College Cork's academic strategy (2018–2022) placed a strong emphasis on such approaches with three of its six pillars being; 'civic and community engagement', 'inter/trans-disciplinarity' and 'sustainability' [51]. While we put out a call for interested participants, we remained clear about the need for individuals with specific experience to come forward. In using the statement below, ten people came forward to participate in interviews within the University:

> potential participants should have experience of working with industry, communities, policymakers, NGOs etc. in a collaborative, engaged, co-produced, and two-way approach e.g. stakeholders are actively involved in research design, in the research work itself, and the implementation of research findings post -project.

The disciplinary bases which were represented through these conversations covered government and politics, ecology, civil engineering, energy engineering, business and marketing, economics, geography, and sociology. This also covered a range of career stages from earlier postdoctoral, up to research fellow, and professorship.

Box 4. Overview of the interviewee's area of research.

Background	No. of participants
Deliberative and participatory governance	1
Energy policy	1
Climate adaptation	2
Community energy planning	1
Lived experience of energy/climate change	2
Food security and sustainability	1
Sustainable agriculture	1
Wildlife conservation	1

Across these conversations three key areas emerged as important insights for both of us as we set out to develop our research in transdisciplinary practice for sustainability. These were: challenges associated with involving a range of different stakeholders in research; institutional challenges emanating from new research practices; and suggested paths forward. Throughout the following sections, the relevant quotes from participants are integrated within the text, with the interview participants (IP) referred to as IP1 to IP10.

2.3 Reflections on Approaches to Transdisciplinarity

While there is currently significant momentum for private, public, and community-based groups to be involved in research with relation to sustainability and climate change, this is not a new concept [64]; "*although it's not as new as you might think, this has been around for decades, but what is new is the current drive to have elements of Interdisciplinary or Transdisciplinary across most of the research" (IP3).* Established participatory research approaches with long traditions of methods, practices, and empirical research are often crowded out by "*the jargon of co-production*" ([58] p. 31). These approaches are not new, but rather there is a new and growing emphasis placed upon them. This creates an issue whereby researchers with no previous experience of these practices or traditions undertake these new roles without recognising this existing knowledge:

> you want to bring in other disciplines but then I suppose you want there to be enough people within a project who understand how this knowledge making works or else you have people reinventing the wheel and suggesting the collaborative community engagement work is some brand new radical type of thing which it isn't it has been around a long time (IP10).

Moreover, there is the question of the variety of different roles researchers must be prepared to fill that is well documented in the literature [18, 21]. An additional point that must be acknowledged is that some people will be more suited to the role than others; "*There is something that no one ever says. The importance of being a nice person is key*" (IP6). Many of the necessary skills for engaging with a range of people are 'soft' interpersonal traits that cannot be developed with conventional training.

It has been claimed that through working with researchers, others are "*rewarded with new knowledge, capacities, and 'voice' by taking part in research networks*" ([65] p. 743). There is a danger here, however, in starting from an assumed position that beneficial outcomes will be achieved simply through the engagement process. This evidence may be influenced by the form of practice being implemented on the ground, whereby a box-tick logic is applied to engagement. Several of the academics involved noted that often within projects working across several disciplines the social science elements (or engagement of different groups more generally) is added on to the physical sciences or engineering rather than being itself embedded within the core of the research process. The risk is that under pressure to work with private, public and community-based groups, these tokenistic inclusions may take place;

"*a project I was involved in previously which had stakeholder engagement aspects... it was lip service... it was just that we have a report on social acceptance here it is, grand, that is that box ticked*" *(IP2)*. Evaluation of engagement activities is thus increasingly important [66].

Clarity on the purpose of working in this manner must be clear for all parties involved. It was noted that there is a requirement for funding to enable researchers and partners time to build trust and solidarity [13]; "*You really try and establish that you are not there with a big agenda to extract information. You are there to show a sense of solidarity as much as anything else and then bit by bit you earn trust*" *(IP7)*. This, however, is something currently not facilitated through typical 2–3-year project timeframes with tight delivery deadlines and narrowly defined goals.

The issue of communication between actors is of crucial importance. Coming from different disciplinary backgrounds or professional settings results in many linguistic styles of specialisation [50]. It can become a difficult challenge to communicate complex findings using simple language [67]. The tension between different knowledge types, technical expertise, organisation structures [68]—and within this language—represents one of the core challenges for research moving from disciplinary to transdisciplinary practices [50, 69].

In the words of Saviano et al., despite the growing importance being placed on transdisciplinarity, "*science has continued to follow a vertical pathway through the development of increasingly specialized knowledge, thus generating fragmentation and divides that make dialogue among disciplines difficult*" ([57] p. 1551). Alongside this, often the proxy variables used in the evaluation of research practices do not include elements related to transdisciplinary approaches [55]. The time required to coordinate meaningful engagement is severely underappreciated. As alluded to, "*what happens often is that the social science component is bolted on*" *(IP10)*. Beyond how knowledge is integrated—and with relation to the definition of roles—time [70] and levels of commitment [61] become important considerations. As noted by one participant; "*one of the things is that people don't understand the time it takes to do this work and the fact you have to develop relationships with people and you have to understand what they want*" *(IP6)*. Researchers must balance the tension between levels of commitment from societal actors, the necessity of producing findings that are practical to these actors and the need for research outputs.

Working in a team that includes a range of different actors, coming from different institutions/organisations, may prove a challenge due to the differing performance expectations and goals within those institutions/organisations [61]. The transdisciplinary team becomes its own entity, but one in which different actors must meet very different performance indicators. Within an academic setting, the importance recently placed upon transdisciplinary approaches to managing complex issues is still at odds with the structural expectations created within academia:

> *when it comes to your own personal career it doesn't really provide … academic outputs will suffer if you are doing this kind of engaged research properly, so I think that has to be reflected. In some respects, people look down... I am not saying that happens across the board... but it is considered lesser possibly than academic research which I would very much argue against* (IP6).

Importantly, as Lauto and Sengoku note with relation to barriers to TdR within universities, career advancement opportunities remain more fixed towards singular disciplinary approaches due to the historical development of disciplinary traditions throughout the twentieth century [54]. As noted within an interview; "*essentially the currency is journal publications… yes, there are lots of scholars that do this kind of work and are held in high esteem, but you need to be publishing journal papers" (IP10).*

This lack of recognition for TdR in career progression was reflected in the ease with which participants could pursue this approach. There was a clear distinction between permanent faculty positions with the freedom to explore interests, in comparison to earlier stage researchers lacking this certainty and security. As noted by a permanent senior staff member; "*I just did my own thing. There was never any issue. The fact of the matter is given my seniority, you kind of work your way into a position of privilege"* (IP7). Linked to this issue of funding is the challenge of maintaining capacity within a research group:

You need to have continuity, but our structures don't allow for that. When I say our, I mean academia more widely. You can recruit as many students as you want, you get in postdocs, and they are there for a period then move on. The knowledge is lost, it's hard to retain capacity in research groups. There are very few, if any, (permanent) places within the research side of it (IP3).

This is of particular importance when it comes to the need for a somewhat unique skillset; such as expertise in sustainability challenges and engagement approaches.

2.4 The Motivating Factors and a Positive Framing

Despite its current difficulties, the significance of senior university management and policy now calling for TdR must be appreciated; "*I think the current senior management in the university value it a bit more than previous ones. So, it is a bit easier to do when you are getting kudos for it from senior university people*" (IP8). For some participants, this is in stark contrast to how they had previously been 'looked down' upon. However, it does not come without concerns, as there are questions over the motives of this new agenda and if it was truly in the spirit of seeking to establish new relationships and knowledge; "*then the other part I suppose is that generally it (host research institute) does recognise this work but it tends to be for social media and that stuff*" (IP6).

Counterbalancing the challenges represented, there were also some clear motivations expressed around engaging with private, public, and community-based groups. Some hints of inspiration for us to follow. "*These issues are wicked problems that mean you have to work across the boundaries" (IP5).* Similar to the findings of Cerrato et al., motivations are not led by an emphasis in funding, but rather facilitated by an emphasis to act upon personal feelings towards what research should seek to achieve in this space [71]. As noted by one participant; "*At the end of the day that is what we are trying to do, make something useful, make a contribution. We are dealing with real life issues that require an immediate response" (IP4).*

Within the established literature intrinsic motivations outlined include; an urge to work towards the improvement of society through contributing to the advancement of the common good, a desire to engage with issues/problems that transcend disciplinary/academic boundaries and a sense of fulfilment based on having an impact in the lives of both the researcher and participants [52]. The potential impact of research within society is an important consideration:

> *You want your research to be used... It is also you want to make some kind of good happen in the world in a way as well. There is those motivation as well. There is very little research done that is not useful. But to actually make it usable you have to cross the divide into practise and seeing... how do we take this useful information and make it useable (IP6).*

The importance placed upon a move to usable knowledge is of central importance with regard to sustainability science and climate action, and can be seen as a key driver in the institutional push. The benefits of working in topic-focused, inter/transdisciplinary orientated research institutes, which can facilitate the bringing together and merging of expertise was noted in some interviews; "*We have had sociologists, architects, geographers, ecologists so it is very mixed TdR membership. That is something that is quite typical in research institutes or environmental institutes. I think that's a strength in terms of promoting TdR" (IP4).* They offer space for the building of relationships, sharing of experiences and collaborative learning; "*I think the structure of having research centres on thematic areas rather than disciplines like we have here... that is a good way forward" (IP5).*

However, one academic highlighted a key challenge in the development of these transdisciplinary orientated research institutes is that they cannot be built on the same foundations as previous research institutes. New (fit for purpose) institutes would require a restructuring of universities to support new ways of coordinating and administrating research:

> *All of our research centres have to belong to a college. You are in here (host research institute), it's one of our flagship institutes it has PIs from across the*

university but yet it must belong to one college. So, it is under the College of Science Engineering and Food Science (SEFS), that's for resource and reporting structure... This then means that academics from outside SEFS are less encouraged to engage with it and contribute because the benefits don't go to their schools or college (IP3).

These new research institutes could help to overcome the structures within academia that hinder progress in the development of transdisciplinary research outlined above: funding procedures, career advancement opportunities, evaluation criteria, etc. [53].

Such a radical re-imagining of university structures will require time. Moreover, perhaps this would be misguided and the best place for these collaborative ventures to take place would be outside of formal structures; "*The nature of academic work now means sometimes it is hard to find a little bit of time for blue-sky thinking and thinking the thoughts that lead to a research question that leads to some output" (IP1).* Ultimately, what is needed for these new, creative ideas to emerge is the space for meaningful discussions to take place and an understanding/relationship to be built. These are far better facilitated through less formal interactions, coming from a place of humility and respect to discuss each other's interests; "*You start talking to all sorts of people across the university and you start conversations with people. Building trust with people that they feel comfortable in a space. These things can't be rushed and they certainly can't be institutionalized" (IP7).* While challenges exist for these forms of research practice, both at the level of the institution and in practice, there are clear intrinsic motivations for researchers committed to TdR.

At a higher level, above how funding within the research institute may be coordinated, there is the question of the funding itself. As recently outlined by Pohl et al., there are three key stages to a TdR project: problem framing, problem analysis and exploring impact [72]. At present, current funding mechanisms generally call for all three of these elements to take place within the confines of a two- or three-year project. It is suggested here that more appropriate calls would involve a staged approach with dedicated strands for exploratory design to build an improved understanding of problems and trust among the stakeholders, as well as much later to evaluate the impact of previous projects. Firstly,

> *Research design is very important. Often, it's the bit left to the last minute and sometimes it's retrospective, on larger studies, it's very rushed. A lot of time is spent doing the research but not enough planning it and conceptualizing what it is you ought to do. There is a saying, measure it twice and cut once, get it right" (IP3). And secondly, "You need to show impact. Environmental, economic, social, technological. That is the requirement. It is not an aspiration. (IP2)*

However, the intervention of a research project is extremely short relative to long-term societal transitions, so impacts are unlikely to be clear until several years after the work takes place.

2.5 Setting Out in a Transdisciplinary Research Journey

Throughout the process of organising and undertaking these conversations on transdisciplinarity, several emergent institutional barriers to co-produced, transdisciplinary research, both across disciplines and with private, public and community-based groups were referenced. This includes barriers to career progression, the rigidity of disciplinary structures, coordination and administration difficulties and the incumbent structure of university schools in relation to establishing thematic research institutes. Within the literature, there is a range of contributions on the institutional challenges associated with transdisciplinary [73, 74] and engaged research [60]. The suggestion can be made that these challenges emanate from the established structures and procedures, particularly with relation to the evaluation of performance within academic institutions. The need for substantial support for researchers working with private, public and community-based groups within departments is noted [75]. Within this, however, difficulties are faced from a cross-disciplinary perspective. Situations in which one department supports the approach while another does not creates power imbalances within the cross-disciplinary structure of the research project.

There is a great risk associated with the mandate across research to pursue these approaches; there are questions to be asked of the motives behind this push, and careful consideration given to the purpose of involving different actors in research. As we reflected upon our own involvement in TdR, it was clear that if you want to do it then you must be "all in". There is a lack of recognition within funding bodies that genuine co-production with communities is very time-consuming and needs to be resourced properly across several fronts (e.g. financially, new skillsets, flexibility in being able to compensate communities, the time required to build trust, etc.). The danger of burnout has been referenced with relation to qualitative research as a whole [76], and as funding continues to prioritise these approaches, research fatigue may increase with the potential to act as a counterproductive force in relation to both climate action and scientific legitimacy [77]. There is a thin line between cultivating co-production and data mining communities to suit the needs of researchers.

The institutional challenge of transdisciplinary research relates to its position within academia. While this position has the advantage of strengthening links at the interface between science and society, it suffers institutionally through a failure to advance theoretical and conceptual developments, coupled with a lack of reflexivity. Transdisciplinary, as such, is often perceived as a 'scientific service' rather than something of genuine interest within established disciplines [63]. Its position within academia must be strengthened at the level of both research and teaching to enhance its development. For practitioners of transdisciplinarity, the benefits and motivations of such an approach must be kept at the forefront of the mind, in light of institutional barriers present. Here, the importance of conversations within academic settings which are "*undertaken in good faith and humility, with the intention of gaining new insights from other disciplinary perspectives and sources of knowledge in order to be better equipped to develop appropriate questions on sustainability related issues*" ([78] p. 237) is a crucial starting point in the journey towards working in a cross-disciplinary manner that can engage with the wider public on the complex challenges of contemporary society.

2.6 Conclusion

The conversations and insights presented through this chapter were convened in the first year of our doctoral research projects. It offered us an important moment to reflect on this path we were taking at an early stage and consider our approach in the wider context of experiences within our host university. One key finding of this process is that there is still a crucial gap between the growing interest versus the reality of practising transdisciplinary research in universities. The emphasis on traditional research outputs and journal publications is often misaligned with TdR processes, however, some improvements on the associated metrics are occurring over time due to increased emphasis on such approaches in funding calls. Several interesting reflections missing from the established literature have been represented through this research process. As two early-career researchers, this process has helped us to understand some of our own experiences of undertaking transdisciplinary research (See: Chapter 9), while exposing us to a broad range of issues related to the topic not directly experienced within our situation. It sought to remind us of the need for disciplinary humility in this process of "journeying together" [79].

We note several existing institutional challenges that hinder progress with the practice of transdisciplinary method. This merits greater investigation into the values and motivations of higher-level institutions that are placing an increasing emphasis on the pursuit of these methods. If transdisciplinary approaches are to be effective, then there is a need to re-imagine the structures of our universities, research institutions, and funding. Considering this institutionalisation of the need for research to work alongside private, public, and community-based groups, a greater understanding of the changing role of practising researchers and associated training and development needs is warranted. For practitioners, awaiting the necessary institutional changes, but driven by a motivation towards working in this manner, an emphasis must be placed on creating space for conversation, adding rigour to practice by highlighting learnings from failure alongside success and building transdisciplinary research networks both within research institutes and more broadly across institutes.

References

1. E.G. Irwin, P.J. Culligan, M. Fischer-Kowalski, K.L. Law, R. Murtugudde, S. Pfirman, Bridging barriers to advance global sustainability. Nat. Sustain. **1**, 324–326 (2018)
2. S.O. Funtowicz, J.R. Ravetz, Science for the post normal age, in *Perspectives on Ecological Integrity*. ed. by L. Westra, J. Lemons (Springer, Dordrecht, 1995), pp.146–161
3. V.A. Brown, J.A. Harris, J.Y. Russell (eds.), *Tackling Wicked Problems Through the Transdisciplinary Imagination* (Earthscan, London, 2010)
4. C. Pohl, P. Krütli, M. Stauffacher, Ten reflective steps for rendering research societally relevant. GAIA-Ecol. Perspect. Sci. Soc. **26**, 43–51 (2017)
5. S. Kampelmann, M. Kaethler, A.V. Hill, Curating complexity: an artful approach for real-world system transitions. Environ. Innov. Soc. Transit. **27**, 59–71 (2018)
6. J. Köhler, F.W. Geels, F. Kern, J. Markard, E. Onsongo, A. Wieczorek, F. Alkemade, F. Avelino, A. Bergek, F. Boons, L. Fünfschilling, An agenda for sustainability transitions research: state of the art and future directions. Environ. Innov. Soc. Transit. **31**, 1–32 (2019)
7. E. Byrne, G. Mullally, C. Sage, *Transdisciplinary perspectives on transitions to sustainability* (Routledge, London, 2017)
8. E.P. Byrne, G. Mullally, Seeing beyond silos: transdisciplinary approaches to education as a means of addressing sustainability issues, in *New Developments in Engineering Education for Sustainable Development*. ed. by W. Leal Filho, S. Nesbit (Springer, Cham, 2016), pp.23–34
9. S.P. Jong, T. Wardenaar, E. Horlings, Exploring the promises of transdisciplinary research: a quantitative study of two climate research programmes. Res. Policy **45**, 1397–1409 (2016)
10. D.W. Cash, S.C. Moser, Linking global and local scales: designing dynamic assessment and management processes. Glob. Environ. Change **10**, 109–120 (2000)
11. R.W. Kates, W.C. Clark, R. Corell, J.M. Hall, C.C. Jaeger, I. Lowe, J.J. McCarthy, H.J. Schellnhuber, B. Bolin, N.M. Dickson, S. Faucheux, Sustainability science. Science **292**, 641–642 (2001)
12. A. Wiek, F. Farioli, K. Fukushi, M. Yarime, Sustainability science: bridging the gap between science and society. Sustain. Sci. **7**, 1–4 (2012)

13. M.C. Lemos, J.C. Arnott, N.M. Ardoin, K. Baja, A.T. Bednarek, A. Dewulf, C. Fieseler, K.A. Goodrich, K. Jagannathan, N. Klenk, To co-produce or not to co-produce. Nat. Sustain. **1**, 722–724 (2018)
14. F. McDermott, Complexity theory, trans-disciplinary working and reflective practice, in *Applying Complexity Theory*. ed. by A. Pycroft, C. Bartollas (Policy Press, 2014), pp.181–198
15. J.H. Spangenberg, Sustainability science: a review, an analysis and some empirical lessons. Environ. Conserv. **38**, 275–287 (2011)
16. K.H. Bieluch, K.P. Bell, M.F. Teisl, L.A. Lindenfeld, J. Leahy, L. Silka, Transdisciplinary research partnerships in sustainability science: an examination of stakeholder participation preferences. Sustain. Sci. **12**, 87–104 (2017)
17. L.G. Horlings, M. Nieto-Romero, S. Pisters, K. Soini, Operationalising transformative sustainability science through place-based research: the role of researchers. Sustain. Sci. **15**, 467–484 (2020)
18. C. Pohl, S. Rist, A. Zimmermann, P. Fry, G.S. Gurung, F. Schneider, C.I. Speranza, B. Kiteme, S. Boillat, E. Serrano, "Researchers" roles in knowledge co-production: experience from sustainability research in Kenya, Switzerland, Bolivia and Nepal. Sci. Public Policy **37**, 267–281 (2010)
19. D. Loorbach, N. Frantzeskaki, W. Thissen, A transition research perspective on governance for sustainability, in *European Research on Sustainable Development*. ed. by C. Jaeger, J. Tàbara, J. Jaeger (Springer, Berlin, 2011), pp.73–89
20. D.J. Lang, A. Wiek, M. Bergmann, M. Stauffacher, P. Martens, P. Moll, M. Swilling, C.J. Thomas, Transdisciplinary research in sustainability science: practice, principles, and challenges. Sustain. Sci. **7**, 25–43 (2012)
21. J.M. Wittmayer, N. Schäpke, Action, research and participation: roles of researchers in sustainability transitions. Sustain. Sci. **9**, 483–496 (2014)
22. H.C. Featherstone, D.L. Owen, Rethinking professional development for public engagement with research: a way to improve uptake and impact of training? Res. All **4**(1), 128–144 (2020)
23. R.A. Pielke Jr., *The Honest Broker: Making Sense of Science in Policy and Politics* (Cambridge University Press, Cambridge, 2007)
24. D. Durant, Are honest brokers good for democracy? Soc. Epistemol. **3**(2), 1–14 (2022)

25. S.P. Turner, Defining a discipline: sociology and its philosophical problems, from its classics to 1945. In: *Philosophy of Anthropology and Sociology* (Elsevier, Amsterdam, 2007) , pp. 3–69
26. D. Mišina, Who now needs sociology? Transdisciplinarity vs. tradition. Can. J. Sociol. **40**, 527–546 (2015)
27. W.K. Carroll, Discipline, field, nexus: re-visioning sociology. Can. Rev. Sociol. Can. Sociol. **50**, 1–26 (2013)
28. A.J. Puddephatt, N. McLaughlin, Critical nexus or pluralist discipline? Institutional ambivalence and the future of Canadian sociology. Can. Rev. Sociol. Can. Sociol. **52**, 310–332 (2015)
29. S. Beder, *The New Engineer: Management and Professional Responsibility in a Changing World* (Macmillan Education AU, London, 1998)
30. E. Conlon, The new engineer: between employability and social responsibility. Eur. J. Eng. Educ. **33**, 151–159 (2008)
31. C. McGookin, B.P.Ó. Gallachóir, E.P. Byrne, *Walking the Walk; Meaningfully Engaging People with Engineering Challenges* (University College Cork, Cork, 2021), pp.1–8
32. M. Yearworth, Sustainability as a 'super-wicked' problem; opportunities and limits for engineering methodology. Intell. Build. Int. **8**, 37–47 (2016)
33. G. Mullally, E. Byrne, C. Sage, Disciplines, perspectives and conversations. In: *Transdisciplinary Perspectives on Transitions to Sustainability* (Routledge, 2017), pp. 21–40
34. J.T. Klein, The transdisciplinary moment (um). Integral Rev. **9** (2013)
35. E. Darian-Smith, P. McCarty, Beyond interdisciplinarity: developing a global transdisciplinary framework. Transcience **7**(2), 1–26 (2016)
36. P. Stock, R.J. Burton, Defining terms for integrated (multi-inter-trans-disciplinary) sustainability research. Sustainability **3**, 1090–1113 (2011)
37. R.J. Lawrence, Housing and health: from interdisciplinary principles to transdisciplinary research and practice. Futures **36**, 487–502 (2004)
38. P.W. Balsiger, Supradisciplinary research practices: history, objectives and rationale. Futures **36**, 407–421 (2004)
39. S.O. Funtowicz, J.R. Ravetz, Science for the post normal age. In: *Perspectives on Ecological Integrity* (Springer, 1995), pp. 146–161
40. M. Stauffacher, A.I. Walter, D.J. Lang, A. Wiek, R.W. Scholz, Learning to research environmental problems from a functional socio-cultural constructivism perspective: the transdisciplinary case study approach. Int. J. Sustain. High. Educ. **7**, 252–275 (2006)

41. R.W. Scholz, D.J. Lang, A. Wiek, A.I. Walter, M. Stauffacher, Transdisciplinary case studies as a means of sustainability learning: Historical framework and theory. Int. J. Sustain. High. Educ. 7, 226–251 (2006)
42. C.M. Raymond, I. Fazey, M.S. Reed, L.C. Stringer, G.M. Robinson, A.C. Evely, Integrating local and scientific knowledge for environmental management. J. Environ. Manag. **91**, 1766–1777 (2010)
43. T.R. Miller, A. Wiek, D. Sarewitz, J. Robinson, L. Olsson, D. Kriebel, D. Loorbach, The future of sustainability science: a solutions-oriented research agenda. Sustain. Sci. **9**, 239–246 (2014)
44. M. Polk, Achieving the promise of transdisciplinarity: a critical exploration of the relationship between transdisciplinary research and societal problem solving. Sustain. Sci. **9**, 439–451 (2014)
45. I. Fazey, N. Schäpke, G. Caniglia, J. Patterson, J. Hultman, B. Van Mierlo, F. Säwe, A. Wiek, J. Wittmayer, P. Aldunce, Ten essentials for action-oriented and second order energy transitions, transformations and climate change research. Energy Res. Soc. Sci. **40**, 54–70 (2018)
46. D. Fam, M. O'Rourke (eds.), *Interdisciplinary and Transdisciplinary Failures: Lessons Learned from Cautionary Tales* (Routledge, Abingdon, 2020)
47. S. Hoffmann, C. Pohl, J.G. Hering, Exploring transdisciplinary integration within a large research programme: empirical lessons from four thematic synthesis processes. Res. Policy **46**, 678–692 (2017)
48. H.J. Aslin, K.L. Blackstock, Now i'm not an expert in anything: challenges in undertaking transdisciplinary inquiries across the social and biophysical sciences. In: *Tackling Wicked Problems Through the Transdisciplinary Imagination* (2010), pp. 117–129
49. P. Woelert, V. Millar, The 'paradox of interdisciplinarity' in Australian research governance. High. Educ. **66**, 755–767 (2013)
50. T. Jahn, M. Bergmann, F. Keil, Transdisciplinarity: between mainstreaming and marginalization. Ecol. Econ. **79**, 1–10 (2012)
51. M. Stauffacher, Beyond neocorporatism? Transdisciplinary case studies as a means for collaborative learning in sustainable development. In: *Environmental Sociology* (Springer, Dordrecht, 2010), pp. 201–216
52. M.H. Guimarães, C. Pohl, O. Bina, M. Varanda, Who is doing inter-and transdisciplinary research, and why? An empirical study of motivations, attitudes, skills, and behaviours. Futures **112**, 102441 (2019)

53. P. Bolger, A study of faculty perceptions and engagement with interdisciplinary research in university sustainability institutes. J. Environ. Stud. Sci. **11**, 115–129 (2021)
54. G. Lauto, S. Sengoku, Perceived incentives to transdisciplinarity in a Japanese university research center. Futures **65**, 136–149 (2015)
55. P. Mårtensson, U. Fors, S.-B. Wallin, U. Zander, G.H. Nilsson, Evaluating research: a multidisciplinary approach to assessing research practice and quality. Res. Policy **45**, 593–603 (2016)
56. A. Ni Shuilleabhain, F. McAuliffe, É. Ní Shé, 'Bottoms up': a case study on integrating public engagement within a university culture. Res. All **5**, 366–381 (2021)
57. M. Saviano, S. Barile, F. Farioli, F. Orecchini, Strengthening the science–policy–industry interface for progressing toward sustainability: a systems thinking view. Sustain. Sci. **14**, 1549–1564 (2019)
58. K.J. Mach, M.C. Lemos, A.M. Meadow, C. Wyborn, N. Klenk, J.C. Arnott, N.M. Ardoin, C. Fieseler, R.H. Moss, L. Nichols, Actionable knowledge and the art of engagement. Curr. Opin. Environ. Sustain. **42**, 30–37 (2020)
59. S. Duncan, S. Oliver, Motivations for engagement. Res. All **1**, 229–233 (2017)
60. T. Augsburg, Becoming transdisciplinary: the emergence of the transdisciplinary individual. World Fut. **70**, 233–247 (2014)
61. W.P. Boon, M.M. Chappin, J. Perenboom, Balancing divergence and convergence in transdisciplinary research teams. Environ Sci Policy **40**, 57–68 (2014)
62. E. Boyle, C. McGookin, C. O'Mahony, P. Bolger, E. Byrne, B.Ó. Gallachóir, G. Mullally, Understanding how institutions may support the development of transdisciplinary approaches to sustainability research, Res. All **7** (2023)
63. I. Hughesis, E. Byrne, G. Mullally, C. Sage, Metaphor, Sustainability, Transformation, (n.d.)
64. E.P. Odum, G.W. Barrett, *Fundamentals of Ecology* (Saunders, Philadelphia, 1971)
65. N. Klenk, K. Meehan, Climate change and transdisciplinary science: problematizing the integration imperative. Environ Sci Policy **54**, 160–167 (2015)

66. M.S. Reed, S. Duncan, P. Manners, D. Pound, L. Armitage, L. Frewer, C. Thorley, B. Frost, A common standard for the evaluation of public engagement with research. Res. All **2**, 143–162 (2018)
67. B. Tress, G. Tres, G. Fry, P. Opdam, *From Landscape Research to Landscape Planning: Aspects of Integration, Education and Application* (Springer, Berlin, 2005)
68. T. Jahn, *"Transdisciplinarity in the Practice of Research", Transdisziplinäre Forschung: Integrative Forschungsprozesse verstehen und bewerten* (Campus Verlag, Frankfurt/Main, 2008)
69. A. Kantamneni, R.L. Winkler, K. Calvert, Incorporating community: opportunities and challenges in community engaged research. Res. Agenda Environ. Manag. (2019)
70. M. Polk, Transdisciplinary co-production: designing and testing a transdisciplinary research framework for societal problem solving. Futures **65**, 110–122 (2015)
71. S. Cerrato, V. Daelli, H. Pertot, O. Puccioni, The public-engaged scientists: motivations, enablers and barriers. Res. All (2018)
72. C. Pohl, J.T. Klein, S. Hoffmann, C. Mitchell, D. Fam, Conceptualising transdisciplinary integration as a multidimensional interactive process. Environ Sci Policy **118**, 18–26 (2021)
73. J.M. Wittmayer, Y.-S.E. Huang, K. Bogner, E. Boyle, K. Hölscher, T. von Wirth, T. Boumans, J. Garst, Y.H. Hendlin, M. Lavanga, Neither right nor wrong? Ethics of collaboration in transformative research for sustainable futures. Humanit. Soc. Sci. Commun. **11**, 1–15 (2024)
74. A.I. Gaziulusoy, C. Ryan, S. McGrail, P. Chandler, P. Twomey, Identifying and addressing challenges faced by transdisciplinary research teams in climate change research. J. Clean. Prod. **123**, 55–64 (2016)
75. A.S. Zimmerman, Community-engaged research: opportunities, challenges, and the necessity of institutional support. Sch. Publ. Res. Methods Discip. IGI Glob. 213–233 (2019)
76. T. Clark, We're over-researched here! Exploring accounts of research fatigue within qualitative research engagements'. Sociology **42**, 953–970 (2008)
77. M.R. Warren, J. Calderón, L.A. Kupscznk, G. Squires, C. Su, Is collaborative, community-engaged scholarship more rigorous than traditional scholarship? On advocacy, bias, and social science research. Urban Educ. **53**, 445–472 (2018)
78. E. Byrne, C. Sage, G. Mullally, Transdisciplinarity within the university: emergent possibilities, opportunities, challenges and constraints. In: *Transdisciplinary Perspectives on Transitions to Sustainability* (2017), pp. 237–243

79. E. Byrne, K. Keohane, A. Revez, E. Boyle, C. McGookin, N. Dunphy, C. O'Neill, Engineering with social sciences and humanities; necessary partnerships in facing contemporary (un) sustainability challenges?, in *Engineering, Social Sciences, and the Humanities: Have Their Conversations Come of Age?*. ed. by S.H. Christensen, A. Buch, E. Conlon, C. Didier, C. Mitcham, M. Murphy (Springer International Publishing, Cham, 2023), pp.375–393

3

The Power of Community: Lessons from Regional Energy Transitions Across Europe

Connor McGookin

3.1 Introduction

There is a growing body of literature on community/local/regional actors pursuing energy self-sufficiency, independence or autonomy with renewable sources [1–4]. Cities, municipalities, regions or islands have great potential to lead the way in the energy transition. Acting locally can help to overcome some of the challenges that transforming our energy system to a fossil fuel free future presents [5–7]. There are several different examples throughout Europe of such leadership: the Covenant of Mayors, 'energy self-sufficient regions' in the European Alps, '2000-W society' in Switzerland, 'Transition Towns' in the United Kingdom, 'bioenergy villages' in Germany, 'energy regions' in Austria, and 'renewable energy islands' in Denmark [8]. The shared goal between all these initiatives is to reduce dependence on fossil fuel imports in place of local renewable sources. For many of these initiatives, a key motivator is that such a transition would help to rejuvenate a rural community that is suffering from decline.

Barriers to the development of renewable energy are primarily non-technical challenges that are dynamic and depend on the context. In the

C. McGookin and E. Boyle, *Engaging Communities in Climate Action*, Energy, Climate and the Environment, https://doi.org/10.1007/978-3-031-93498-8_3

case of opposition to wind energy, existing research has shown a complex variety of conditions that shape public perception including physical, contextual, political, economic, social, local and personal aspects [9–13]. A key factor behind these issues is the perceived lack of fairness and distrust, as the public feel excluded from planning/decision-making processes [13–15]. Community energy projects offer a means to address some of these tensions by providing mechanisms for community ownership and more open decision-making processes [16–18].

Given that regional energy transitions are generally situated in small communities, many relevant actors already know each other from frequent face-to-face interactions [19]. Participants in a study examining two German case studies emphasised the importance of the informal, personal networks during the evolution of the local energy transition [6]. Through these networks more effective lines of communication can be formed which facilitate the creation of a more transparent decision-making process. An essential part of community acceptance is procedural fairness during the placement of technologies. This may mean choosing less economically favourable solutions in the interest of broader community support [20].

Local mobilisation is encouraged by strategically developing storylines around economic development such as the objective of creating regional value by reducing dependence on the import of fossil fuels [19]. The ambition of improved energy security motivates a broader audience than environmental concerns [2]. This is prevalent throughout exemplar areas explored in this chapter such as Güssing, Austria and Samsø Island, Denmark. Suffering from economic crises, made the opportunity for new employment and a reduction in imported fossil fuels in favour of local resources a strong unifying vision for the community.

This chapter will examine four case studies from across Europe to establish what prompted the communities to pursue an energy system transformation, what renewable technologies were developed, the impacts that the projects had on the communities, and finally, what lessons can be taken from these examples to inform efforts elsewhere. We will see that the regions are comparable but yet also unique. Each community works within their own context to harness the resources available to them. The choice of renewables reflects the geography of the regions, and highlights that there is no single solution but rather a variety of options available to us to try and untangle the complex web that is our energy system's reliance on fossil fuels. However, there are also common threads which emerge such as the defiance of isolated rural communities and their fight to thrive.

The case studies are predominantly rural areas, which was not so much by design but rather due to the fact that the pioneering examples happen to be rural regions. Renewables are more likely to be situated in rural landscapes, and thus, this is perhaps to be expected. In addition, as we shall explore in this chapter, the isolation experienced by remote rural areas is a key context within which these initiatives emerge. However, it is important to note that there are also many interesting examples of city's demonstrating leadership [21–23]. Networks like C40, Carbon Neutral City Alliance or ICLEI–Local Governments for Sustainability are driving climate mitigation and other sustainability goals. Through C40, over 700 cities have committed to halve emissions by 2030 and reach net zero by 2050 [24].

3.2 Austrian Energy Regions, Güssing

The Austrian town of Güssing is a flagship example of regional energy transformation. Efforts to transition to an energy system based on local renewable sources started there as early as 1990. It was the first European Union (EU) community to successfully move to 100% renewable energy for electricity and heating/cooling.

3.2.1 Background

The small town of Güssing (~4,000) is the capital of the Güssing region, which is the largest district in the Southeast region of the Burgenland. The wider region has a population of around 30,000, and covers an area of 485 km^2 [25]. It is ~ 48% agricultural land and ~ 42% forest [8]. During the 1980s, this region was one of the poorest in Austria [26]. Having suffered 50 years bordering the iron curtain on the "Eastern bloc" along the Hungarian border, major trade or industrial businesses did not exist in the region and there was a serious lack of transportation infrastructure. This resulted in a scarcity of jobs, with 70% weekly commuters (predominately travelling to Vienna, 160 km away), and a high rate of migration to other regions [27].

A substantial amount of money left the region through imported energy (oil, electricity, etc.), while existing forestry resources remained largely unused. At the end of the 1980s, three friends in Güssing recognised the situation and decided to undertake measures to revert the negative trend; they were then Town Council member, Peter Vadasz, electrical engineer, Reinhard Koch and Herbert Sattler, head of the local timber growers association [26]. On examining the Güssing budget, which repeatedly struggled with a negative balance sheet, they found that most of the money was to cover the costs of energy. In 1988, Koch prepared a report for the council showing that the town of Güssing alone was spending €6 million on imported energy [27]. This led to the 1990 resolution from the Town Council that all public buildings would change their heating systems to use renewables and become 100% fossil fuel free [26]. In 1992, as the newly elected mayor, Peter Vadasz appointed Reinhard Koch to assess how the town could benefit from its natural resources, and to handle the management of the whole project [27].

3.2.2 Regional Energy Transformation

Early examples of energy projects in Güssing were led by the Town Council. To achieve the goal that all public buildings would change to renewable powered heating systems and become fossil fuel free, energy efficiency measures (such as better windows and insulation) were undertaken in all public buildings and public street lighting was upgraded. Within two years, this lead to a 50% reduction in energy demand [26].

In 1991, the Güssing rapeseed-mill went into service to produce biodiesel. Koch's analysis had shown that transportation was the 2nd highest energy demand at 30%, after heating at 50%. The plant had a capacity to produce 840,000L biodiesel per year from local rapeseed [27]. However, producing biofuels in this manner with dedicated energy crops (known as first-generation biofuels) lost popularity due to the competition for land with food. With EU policy restricting the use of first-generation biofuels, the plant was eventually closed in 2005.

In 1996, following Austria's entrance into the EU, several grants became available. A 3MW wood-chip boiler and district heating network was developed, with a yearly output of 45,500MWh. Work on the DH system began by digging up the town's streets to take the insulated pipes carrying the hot water through the whole town. This included homes that had not signed up for the supply, so they could join later if they wished. Some residents were unconvinced, but later realised the benefits when the DH system was reported as being 30% cheaper than existing boiler systems [27].

Turning attention to electrical supply, a combined heat and power (CHP) plant running on biogas produced from the gasification of wood was developed. At the time, gasification was a relatively unproven concept, but through collaboration with scientist, Hermann Hofbauer and Vienna's Technical University a new process was tested [28–30]. Unlike the typical gasification plants around the world, this type of plant avoids the build-up of tar when wood is heated. The process involves producing natural gas by using steam to separate carbon and hydrogen from scrap lumber, and then recombining the molecules. The natural gas fuels an engine that produces electricity and the by-product heat is used to produce warm water for the DH system. The plant, which started operation in 2001 has a rated fuel capacity of 8MW and produces 2,000kWh electric power as well as 4,500kWh heat for district heating at a feed rate of 2,300kg wood per hour [25]. Two parquet flooring companies settled in the region (in 2001/2002), which provided a valuable new stream of sawdust for a second cogeneration plant [8]. The two CHP plants were then able to provide all the electricity demand of the people living in the town of Güssing, and together with the wood-chip boiler, ~ 95% of the towns heat demand was now covered by biomass[28] (Table 3.1).

Following the success in the town of Güssing, ökoEnergieland was formed by an association of 14 surrounding municipalities. They were interested in learning from the experience of the town and expanding the regional energy vision to the wider area. The district heating system

Table 3.1 Renewable energy developed in the town of Güssing [25]

Plant	Source	Capacity
CHP (combined heat and power plant); steam turbine with heat network	Saw dust	8.6 MW fuel capacity, 1.7 MW electrical capacity and 3.5 MW thermal capacity
CHP; wood gasification, R&D, heat network	Wood chips	8 MW fuel capacity, 2 MW electrical capacity and 4.5 MW thermal capacity
Heat plant with heat network	Wood chips, waste wood	17 MW (only heat)
PV (photovoltaic cells) installation + solar boilers	Solar	10 kW peak electrical capacity and 40m^2 solar thermal panels
PV installation	Solar	27.9 kW peak electrical capacity
Heat plant + solar boilers with heat network	Wood chips, solar	650 kW + 320 m^2 solar thermal panels (only heat)
Heat plant with heat network	Wood chips	300 kW (only heat)

was expanded with the installation of a 5 MW boiler in 1998, and several other extensions followed, resulting in a 42 MW capacity system along a 27 km network. While the town had managed to meet all its electricity and heating requirements with biomass, the wider region looked to utilise other technologies as well, with solar energy, geothermal and hydropower being developed [8].

3.2.3 Impact

In the Town of Güssing the results helped the development of the town [31], with:

- 50 New companies
- Over 1,100 new jobs
- Net municipality income of > €9 million/year

The achievements of Güssing are internationally recognised, with many prizes awarded:

- Energy Globe Austria 2005
- Climate Alliance Award 2004
- Austrian and European Solar Prizes 2004 from Eurosolar (as a model region)
- Most Innovative Municipality in Austria 2004
- Most Environmental-Friendly Municipality in Austria 2002

The European Centre for Renewable Energy was established in Güssing in 1996. It was a co-founder of the ökoEnergieland and organises tours through the ökoEnergieland as well as lectures and training courses. Following the success in Güssing, the Climate and Energy Model Regions programme under the Climate and Energy Fund was launched in 2007 by the Austrian federal government to support regions that set the ambition to "become independent of fossil fuels" [32]. In 2023, the network had grown to 124 Climate and Energy Model Regions with 1,134 participating communities [33].

3.2.4 Lessons Learnt

Güssing's success stems from the visionary leadership shown by the mayor (Peter Vadasz), the town council, and electrical engineer (Reinhard Koch) [27]. However, they could not have done it alone. Scientists, engineers, entrepreneurs, teachers and cultural activists also had a role to play at different stages of the transition process [8]. The leadership of the local government was clearly an important starting point, but to turn a plan into action, researchers and entrepreneurs are needed. Through collaboration with researchers, entrepreneurs were able to support the development of new renewable energy technologies and establish Güssing as a research and development hub. This led to the establishment of a coordinating institution, the European Centre for Renewable Energy. In later stages of the transition process, educational institutions and the tourism sector played an important role in creating a regional identity around the energy region vision, and also, building wider public support for the renewable energy projects.

The project was not without its tensions. According to a study on the landscape impact, interviewees in Güssing reported that the impact of renewable energy technologies siting was not considered until problems arose [34]. Soon after the opening of the heat plant and the heat network in Güssing, the chipping of the wood caused a noise and dust nuisance for the neighbouring school, which led to a 'massive protest'. However, according to one interviewee: "*That has been ended very quickly by the municipality. The operators of the district heat plant and the school management agreed that the chipping should take place in the forest instead of the plant*" [34, p. 9]. This shows how quick intervention and open discussion can resolve planning disputes.

3.3 Denmark's Energy Island, Samsø

Samsø 'energy island' is another well-known example of regional energy transformation. Stemming from a funding competition run by the national government, it highlights how public investment can kickstart important societal goals.

3.3.1 Background

In 1997, the Danish Ministry of Environment and Energy sponsored a renewable energy contest challenging communities to abandon fossil fuel dependency, as part of the Danish Energy Action Plan of 1996, Energi21. Interested communities had to submit a master plan with a summary of current energy balance and ways to meet their needs with renewable energy. There were three key priorities in the competition: 1) the reduction of energy demand across sectors (heating, electricity and transportation), 2) local participation in the project, and 3) description of how the region would act as a demonstration project to showcase renewable energy [35].

Samsø was chosen from five competing islands, with a proposal that the island could power and fuel itself entirely on renewable energy within a ten-year period. Interestingly, an engineer who didn't actually live on the island developed this plan. Samsø is a small island in the Kattegat region, 15 kilometres off the Jutland Peninsula. It had ~ 4,300 inhabitants in 1998 living on the 114 km^2 area. While the competition didn't provide support for the renewable energy technologies, it did provide the salary for local, Søren Hermansen, tasked with delivering it [36]. Hermansen had recently leased his fields to a neighbour to take up a job teaching environmental studies at a local boarding school. As reported in the New Yorker, after taking over the family's farm when his father retired, he discovered he wasn't suited to it; "*I like to talk, and vegetables don't respond*" [37].

3.3.2 Regional Energy Transformation

The Samsø Energy Company (SEK) was set up in 1998 with the mission of implementing renewable energy projects that had been identified in the masterplan: wind energy and district heating [35]. Hermansen's role and local knowledge proved to be critical in mobilising the community and conveying their concerns. The proposal was initially met with a lot of scepticism, primarily due to the financial cost. However, when Hermansen proposed the project as a social undertaking, asking community members to think about what it would be like to work on it together and have something to be proud of, then the initiative began to gather interest. With increasing success and popularity, seminars on wind energy were held and soon energy development cooperatives began to emerge [37].

Wind energy was a key focus of the project [35]. In 2000, 11 one-megawatt (MW) wind turbines supplied the island's 22 villages with enough electricity to make it self-sufficient. The placement of the turbines was decided by the locals, planned democratically in order to ensure everyone in the community was happy. An additional 10 offshore wind turbines were later erected in 2002, generating 23MW of electricity. These are considered to offset emissions from the island's cars, buses, tractors and ferries that connect it to the mainland. The Municipality of Samsø financed five of the turbines; 3 are "commercial" owned by larger investors, and the last 2 are owned by smaller shareholders, altogether about 1,500 shareholders organised in two separate companies.

Following the success with wind energy, several small district heating networks were developed to tackle the challenge of renewable heating. Prior to the project, the village of Tranebjerg had already opened a straw burning district heating system back in 1994. In 2002, the district heating plant located between the villages Mårup and Nordby opened. Here, the heat is produced by burning wood chips (about 80%), and by a 2,500 m^2 solar thermal system (about 20%) (Table 3.2). In 2003, a shredded straw burning plant opened to supply the district heating for Onsbjerg. In April 2004, the 'Brundby-Ballen District Heating Cooperative' initiated construction and the islands forth district heating plant opened in 2005. Plans for another district heating plant for some of the remaining villages were unsuccessful, as residents had already invested in replacing oil boilers with biomass stoves and other renewable energy systems such as solar thermal.

While there was no dedicated funding for the renewable energy projects on Samsø, they availed of the Danish Energy Authority subsidy programmes available nationwide. Between 1997 and 2001, solar thermal panels received a subsidy of up to 30%, biomass boilers up to 20%, and heat pump units up to 15%. In both 1999 and 2000, the Danish Energy Authority granted the RE-island project the necessary funds to hire independent energy consultants who could offer private homeowners living outside the proposed district heating areas a free

Table 3.2 Renewable energy developed in Samsø [35]

Plant	Source	Capacity
11 onshore wind turbines	Wind	1 MW each (electricity)
10 offshore wind turbines	Wind	2.5 MW each (electricity)
3 × Straw heat plants with district heat network	Straw	3 MW (heat) + 0.8 MW (heat) + 1.6 MW (heat)
Wood chip heat plant + solar thermal with district heat network	Wood chips and Solar	1.6 MW (heat) (2500 m^2)

energy appraisal of their home. All three district heating systems received grants from the Danish Energy Authority's 'From the Ground Up' fund.

3.3.3 Impact

Thanks to the district heating networks, between 1997 and 2005, the share of renewable energy in heating on Samsø increased from 25% to 65% [35]. There was also a 10% decrease in total heat demand. In contrast, electricity demand has remained the same because despite improvements like LED lighting and energy saving practices, household electricity demand increased over that period.

For electricity, Samsø island now produces significantly more renewable electricity than it needs. In 2013, they exported 70% of the electricity generated [38]. It is also more than the energy demand from transport in diesel and petrol. Since they produce more renewable energy than their total energy demand the island claims to be carbon negative, with a carbon footprint of -2.3 tonnes CO_2 per capita in 2009 (ibid), compared to an EU average of 7.1 tonnes CO_2 per capita [39]. It is estimated that investments in the island totalled €57 million between 1998 and 2007 [35].

Awards:

- UN Global Climate Action Award 2021
- RESponsible Island prize from the EU 2021
- In 2009, Søren Hermansen was named one of the winners of the Göteborg Award for Sustainable Development

Samsø is a very popular tourist destination. It receives around 300,000 visitors a year [40]. Prior to the renewable energy projects, this was driven by Samsø's natural beauty and island culture. However, it has since expanded into educational tourism, with an additional 10,000 energy tourists a year. Between 2005 and 2006, the Samsø Energy Academy was built by local carpenters as a demonstration building [35]. It opened in 2007 as a centre for education and training on sustainable energy technologies, as well as sharing the Samsø island experience. They developed the 'pioneer guide' to share their story and highlight to other communities what they can do to follow suit [41].

3.3.4 Lessons Learnt

For siting the heat plants and wind turbines, visibility, as well as potential noise and dust nuisance played an important role. Locations were proposed, for example, by the Samsø Energy Company and then discussed with the inhabitants and other groups, such as the municipality. By organising an open planning process, consensus was reached without having to compromise restricted areas, such as the conservation area in the north of the island or areas that have many (summer) houses [34]. This along with the financial participation by the inhabitants, in the form of the shared ownership of the wind turbines, was crucial, because there was little external funding. An interesting note is the below quote from Hermansen about the importance of ownership [37]:

> *If I'm reduced to being a customer, then if I like something I buy it, and if I don't like it I don't buy it. But I don't care about the production. We care about the production, because we own the wind turbines. Every time they turn around, it means money in the bank. And, being part of it, we also feel responsible.*

There was some discontent at the fact that the earnings were not shared evenly. Not everyone could afford to invest in the energy projects and thus were excluded from the profits. This left some resenting the wind turbines. However, the commitment from the developers not to pursue the most economically favourable sites but rather those most popular with the community still ensured that people were satisfied that the decision-making process was fair [20].

3.4 German Bioenergy Villages, Jühnde

Jühnde (Lower Saxony) was Germany's first village to produce heat and electricity by means of renewable biomass. Like Samsø, the project originated from a competition but in this case, it was driven by the University of Göttingen as opposed to the national government. It thus poses interesting questions on the role of research and how universities can work in partnership with communities to address critical challenges such as climate and energy.

3.4.1 Background

In 1998, a transdisciplinary team of researchers in the University of Göttingen from seven disciplines drafted a research proposal for a pilot 'biovillage' [42]. It was a diverse interdisciplinary team of sociologists, psychologists, political scientists, economists, agronomists, agrarian economists, biologists and geologists. Though initially ignored, in 2000 the German Ministry for Food, Agriculture and Consumer Protection (BMELV) accepted the proposal as it appreciated the project's potential to provide sustainable rural development.

The research team initially identified 54 potential village partners in the region. From this, 17 villages volunteered to become the 'Bioenergy Village' of which, four villages were chosen for a feasibility study. The research team looked for a village community with motivated, qualified persons, and a village environment with agricultural land [43]. From these four, Jühnde, located 12 km southwest of Göttingen, was selected as the model village as it had the best prerequisites for the transformation into a bioenergy village [42]. The village of Jühnde represents a small part of Lower Saxony, at ~25 km^2 and with ~ 800 inhabitants in 2002.

3.4.2 Regional Energy Transformation

In May 2002, the 'Bioenergy Village' cooperative was founded and established membership contracts with some 70% of the Jühnde inhabitants [43]. Financial support was made available from the national and the regional governments. Over the next 2 years, 2002–2004, preparations were undertaken to install the necessary infrastructure. The participation of locals was central to the approach taken. Residents were not only responsible for planning the details of the project but also for the installations themselves [42].

After the initial general meetings with all the villagers, eight working groups were formed. In these working groups, several relevant project aspects, which the university's team proposed and initially moderated, were discussed: agricultural resources, electricity production, heat production, the heat distribution grid, and the form that the company was to be founded would take. From these groups a central planning (decision-making body) was formed comprising of heads of the specific planning groups and representation from the local authorities. This created a very open, transparent planning process at different levels within (1) the specific planning groups, and (2) the central planning group, as well as (3) the regular inhabitants' meetings.

The renewable energy plants that would provide electricity and heat in Jühnde were installed between 2004 and 2005 (Table 3.3). It consists of three main components [42]:

1. A combined heat and power (CHP) generator with an electric capacity of 680kW$_e$ and roughly 970kW$_{th}$ space heat by burning biogas. The biogas comes from an anaerobic digestion plant with two fermentation units. It digests liquid manure (about 10,000m^3/year) and crops cultivated on approximately 220ha of arable land around Jühnde. The electricity from the CHP unit is fed into the national electricity grid, with a guaranteed feed-in price of €0.17/kWh in 2004 for 20 years. The heat output is mostly used for space heating, meeting about 75% of the village households' hot water demand, and the digestion process. In summer, surplus heat is used to dry wood chips and cereals.
2. To cover peak winter demand, a wood chips furnace with a thermal capacity of 550kW was also installed. In addition, an oil-fired boiler with a capacity of 1,600kW has been installed as a backup. It is reported to be used for less than 5% of the annual heat demand, highlighting the reliability of the system.

Table 3.3 Renewable energy developed in Jühnde [44]

Plant	Source	Capacity
Biogas installation (CHP) with heat network	Manure, energy crops	700 kW
Heat plant	Wood chips	550 kW
PV installations	Solar	(unknown)
Biogas installation (CHP) with heat network	Manure, energy crops	250 kW
Heat plant	Biomass	500 kW
PV installation	Solar	30 kW

3. The district heating system consists of a 5.5km long hot water grid, which delivers steam to the connected households in the village. In the houses, boilers were replaced with heat exchangers.

3.4.3 Impact

Since 2006, the village has been able to cover all its heat demand with bioenergy and produces twice as much electricity as is needed (Fig. 3.1). As a result, the CO_2 emissions per capita were reported as 60% lower than the German average in 2006 (4.3 CO_2 tonnes/year compared to 10.4 tonnes CO_2/year) [45]. The new system also brought economic benefits estimated to save household in Jühnde about €1,000/year [46].

The Agency for Renewable Resources now lists over 180 bioenergy villages across Germany [47]. They define a bioenergy village or community as a region that covers at least 50% of its electricity and heat from locally sourced bioenergy. Thanks to these efforts, bioenergy provided 15.6% of Germany's heat demand in 2023 (170.6 TWh out of 1,093 TWh) and 9.3% of the electricity (49.3TWh out of 525.9 TWh) [48]. While this progress is to be celebrated, we are reminded of the scale of the challenge that decarbonisation poses.

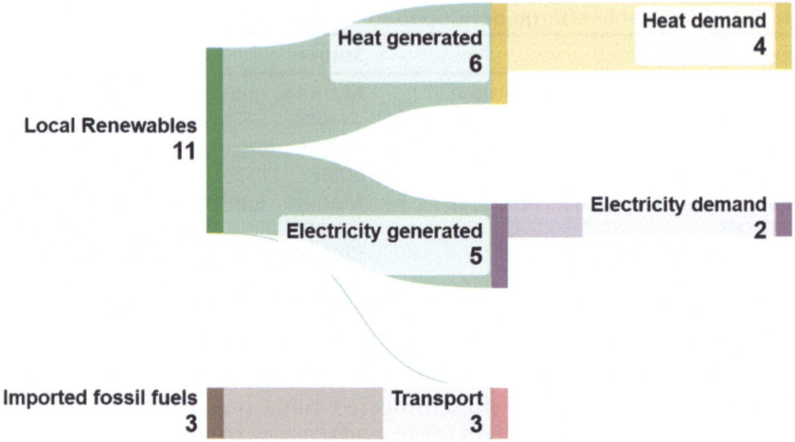

Fig. 3.1 Jühnde energy balance (values in GWh) [46]

3.4.4 Lessons Learnt

As initiator of the Bioenergy Village model, the Interdisciplinary Centre for Sustainable Development (IZNE) project group from Göttingen and Kassel University was critical in defining the concept and prompting the transition to renewable energy sources. It supervised the entire process, conducted progress assessments, and motivated community members to become part of the process [36]. This demonstrates the impact research can have when pursuing an action-oriented, engaged research or transdisciplinary approach.

The university team organised a large variety of activities, ranging from planning workshops to social activities, for the residents to get involved. The initial engagements focused on providing information on the benefits of transitioning to a bioenergy system and becoming 'prosumers' (providers of energy as well as consumers). For planning, there were many village meetings and workshops. In addition, social activities were organised, such as festivals and contests [49]. To manage projects, working groups were established around key areas such as 'Operating company', 'Biogas facility', 'Biomass energy plants', 'Heating plant', 'District heating system' and 'Building energy systems' [43].

This community centred approach and clear decision-making structure ensured a smooth delivery of the project.

3.5 Scotland's Energy Islands, Orkney

It is a little-known fact that the global centre for marine renewable energy research lies just off the North Coast of Scotland on an archipelago of small islands. The energy islands of Orkney are home to some of the fiercest tides and currents in the world. This, coupled with the fact that it is connected to the national grid, makes it the ideal location for testing wave and tidal energy devices. Over the last two decades, the unique community spirit of 'Orkney Ltd.' has been working on a new energy future despite the many challenges this isolated and weathered area faces.

3.5.1 Background

Orkney is an archipelago of 70 islands (20 of which are inhabited) located 16 km north of the Scottish mainland [50]. The majority of the 22,400 population live on the largest island, Mainland Orkney [51]. Orkney has a rich Viking heritage and is renowned for its prehistoric sites [50]. These include Skara Brae, Europe's most well-preserved Neolithic village, and the Ring of Brodgar and Stones of Stenness, Neolithic stone monuments older than Stonehenge.

Similar to other remote communities, Orkney was dependent on expensive imported fossil fuels to meet its heating and transport needs, while the majority of electricity came from the subsea power cable connecting Orkney to the UK national grid [52]. This combined with a harsh climate, poor housing and low wages meant the region had one of the highest levels of fuel poverty in Scotland [53]. The Orkney Renewable Energy Forum (OREF) was created in 2000 with the combined goal of reducing energy demand while also increasing the proportion of Orkney's energy supply that comes from renewable sources [54].

3.5.2 Regional Energy Transformation

Between 2000 and 2012, over 30 MW of wind energy was installed in Orkney [55]. This meant that by 2013, Orkney became a net exporter of electricity, producing 103% of its total electrical needs from renewable sources [53], which has since rose to ~ 130% [55]. With support from Orkney Islands Council, Highlands and Islands Enterprise and Heriot-Watt University (ICIT), the European Marine Energy Centre (EMEC) was opened in Stromness in 2004 with two test sites (wave and tidal) able to accommodate many different designs [56]. This saw the first grid connected wave energy device operational in 2006. There has been a lot of interest in further expansion with large amounts of wave and tidal installations, but these developments have been stalled as the electricity grid cannot accommodate any new generation. Since 2012, Scottish Hydro Electric Power Distribution (SHEPD) have imposed a moratorium on all new generation in Orkney due to increasing constraints on the grid [55].

A key local priority has always been to reinforce Orkney's local network and increase the capacity of the subsea cables between Orkney and mainland Scotland [50]. However, progress has been slow to get the necessary capital investment from the UK government. In 2009, SHEPD upgraded Orkney's local network to become the United Kingdom's first smart grid using the new Active Network Management (ANM) approach [57]. This system manages the energy flow on the grid to reduce curtailment. It was later upgraded in 2013 with the addition of an energy storage system (ESS) by installing a 2 MW/1MWh lithium-ion battery [57].

To overcome the grid limitations, the Orkney Island Council published a hydrogen strategy in 2020 [58]. It looks to expand on the existing projects: Surf'n'Turf and BIG HIT (Building Innovative Green Hydrogen systems in an Isolated Territory). There are two electrolysers to produce hydrogen from the energy generated by the tidal turbines at the EMEC test site in Eday, and the 900kW community-owned wind turbines on the islands of Shapinsay and Eday. The produced hydrogen is stored in fuel cells, which are utilised to provide heating for Shapinsay School, power to some of the inter-island ferries that dock overnight

Table 3.4 Renewable energy developed in Orkney [55]

Plant	Source	Capacity
Wind turbines	Wind	44.7 MW
PV installations	Solar	1.4 MW
Wave and tidal devices	Ocean currents	2 MW*
Hydrogen Electrolysers	Curtailed electricity	1.5 MW

* The output of the EMEC test site varies but it generally has 1–2 MW prototypes installed

at Kirkwall harbour, and to supply the Hydrogen Refuelling Station in Hatston (Table 3.4).

3.5.3 Impact

The marine renewable energy industry represents a significant part of the island economy with around 300 people working in it [50]. There are thirty island companies across the supply chain, sea and onshore surveying, and environmental consultancy. The islands produce more electricity from renewables than they need. However, progress in other areas (transport and heating) has been limited. Emissions declined from 268.7 ktCO_2 in 2005 to 192.5 ktCO_2 in 2017, a 28% decrease [55].

3.5.4 Lessons Learnt

Isolated communities must look out for themselves more so than areas that are closer to the seats of power in national government. This instils a sense of collective purpose. Those working in the marine energy industry in Orkney, do not see themselves simply as private companies, but instead as part of a wider community effort to ensure there is a long-term future for people living there. This is highlighted by the example Laura Watts shares in her book when Gareth Davis, CEO of Aquaterra, presents to an interested venture capitalist the potential to work with the islands as a whole, often referred to as 'Orkney Ltd.', rather than what his business can offer. This collective *"Orkney Ltd. is defined neither by business interest nor industry sector. Its remit and borders are geographic. All those who live within the archipelago are included in its development and future*" [50, p. 197].

3.6 Key Learnings

Regional energy transitions are diverse. They can come in a wide variety of different sizes, can be led by different actors and involve different renewable energy technologies. This diversity is precisely the strength of acting local. One size doesn't fit all. The energy transition needs to be rooted in the local context and engage those impacted. However, there are still some common threads that can be drawn from the case studies explored in this chapter.

3.6.1 What Are the Factors of Success?

The Power of the Guiding Vision

The most obvious common thread throughout the case studies discussed in this chapter, and indeed others that could not be included, is the power of a unifying vision. For remote rural areas that are suffering from social and economic challenges, the goal of energy independence is a strong message to get behind. Communities on the edge have always had to look out for themselves. The regional energy transition is just another in a long tradition of social undertakings that these communities must do to survive. The objective of replacing costly fossil fuel imports with local renewables taps into regional pride.

It's a Social Undertaking

With the guiding vision in place, the regional energy transition spreads through the community and begins to gather interest. The economic incentive of reduced fuel imports is a strong motivator. However, another important element that should not be overlooked is that this is a social undertaking, a way to bring the community together and work towards a shared goal. In most cases, complete community ownership of projects will not be feasible as the amount of capital needed is simply too high. However, people can still be given a sense of ownership in the projects through open and transparent decision-making processes or working on the construction and development. This is exemplified by the example of Jühnde, where the laying of the district heating pipes was a community project that the volunteers got involved with to dig the trenches. Dóci & Vasileiadou from their interviews of participants note that volunteering presented "*the opportunity to get to know each other and to gain new friendships*" [59, p. 9], which was particularly important for newcomers in the community.

Importance of Local Champions

A clear factor for success is the dedicated local champions. In Jühnde, the mayor was an important promoter who motivated the inhabitants to support the idea and to take part in the project. In Güssing and Samsø the projects benefitted greatly from having Reinhard Koch and Søren Hermansen to coordinate projects. There is a huge amount of work that goes into planning a renewable energy project from initial feasibility to planning to operation. While cooperative ownership models offer a great way to ensure benefits can be shared locally, volunteers cannot be expected to deliver projects all on their own. Project management roles are initial to drive initiatives forward.

An important benefit of having these local champions is their existing network within the community and familiarity with the area. As noted by Mundaca et al. in the case of Samsø: "*Hermansen's role and local knowledge proved to be critical in mobilising the community*" [20, p.5]. He was able to meet with key community figures before formal consultations took place for proposed developments and convey their concerns back to the developers. This is noted as an important mechanism by which tensions were avoided.

It's a Collaborative Effort

Local champions are needed to coordinate and drive the regional energy transition process but they cannot do it alone. These are multi-stakeholder efforts, which require collaboration across the community. Energy is entangled in everything we do; thus everyone has a part to play in transforming the energy system. A collaborative approach is essential for wider societal support. This means following collaborative engagement principles when planning for the renewable energy projects. Working with a diverse group of actors to deliver the projects. Forming new partnerships to share expertise and working together towards the community's shared goal.

Establish a Governance Structure

Another important component is the forming of a governance structure. There needs to be a clear decision-making process and coordinating body for the projects. This should involve a mix of different representative voices from the community. The working groups in Jühnde are a good example of this. Although the project was initiated by the university, once the village was chosen to be the model region, working groups were formed with members of the community, and these then fed back into an overarching steering group formed with key representatives.

Identify the Best Renewable Resource Available, There Is no Single Solution

As noted in the Introduction of this chapter, unlike fossil fuels that can be transported to anywhere in the world, renewable energy relies more on the geography of an area. We can see from the four case studies explored here that while there are similarities in what factors drive success, the renewable energy technologies pursued will differ. The town of Güssing made use of the vast forestry resource surrounding it, Jühnde took advantage of the agricultural land, while the islands of Samsø and Orkney look to capture the power of the wind and sea battering against their doorsteps. The move to a renewable energy system represents a shift to a more diverse and decentralised system that must be designed with more careful consideration of the land it will inhabit and the people it will serve.

3.6.2 What Are the Areas for Improvement?

Transport Remains a Difficult Sector to Decarbonise

Güssing claims energy self-sufficiency, because the region generates more renewable energy than their total energy demand. However, in reality, transportation still relies on fossil fuels. The biodiesel plant that opened in 1991 produced from locally-grown rapeseed, a first-generation biofuel, due to a change in the EU biofuel policy and technology improvement, had to close in 2005 [27]. The generation of third generation biodiesel from wood chips opened in 2004, in an experimental setting near the newest CHP plant, but was reportedly bankrupt in 2008 [8]. Likewise, Samsø claims to use 100% renewable energy because the vast majority of offshore wind generated is exported and covers the total energy demand of the island. However, an assessment in 2015 found that the island actually had a renewable share of 60% as there remained some oil boilers in isolated buildings not connected to the district heating network and transport was still almost entirely fossil fuels [60]. Germany's bioenergy villages also focus on electricity and heat [47], and in Orkney, transport remains reliant on oil [55]. There is clearly still work to be done to ensure the transport needs of dispersed rural communities can be met without fossil fuels.

The Importance of Emissions Accounting

The guiding vision of energy independence or 100% renewable energy is undoubtedly a strong unifying message for communities. However, there is a risk that it distracts from the core problem: CO_2 emissions. By examining energy flows, regions claim to be 100% renewable or self-sufficient as their exports of renewables are greater than the amount of fossil fuels imported. This does not, however, reflect a fully decarbonised system. Samsø, despite claiming energy self-sufficiency, still emitted 28.5ktCO_2 emissions in 2015 [60]. That equated to 7.7tCO_2/capita, which is significantly higher than the -2tCO_2/capita previously claimed based on the emissions saved by offshore wind exports. Similarly, in Orkney, despite all the progress with renewable electricity, emissions per capita were 8.7tCO_2/resident in 2017, representing the 5th highest per capita emissions out of all 32 council areas in Scotland [55]. This highlights just how difficult decarbonisation is for dispersed or remote rural populations, which generally have higher energy demand for heating and transport than urban areas.

Another issue in this regard is how running on 100% renewable energy is framed. A region may produce enough electricity to theoretically cover the total electricity demand. However, this only represents about 20% of the energy system. Moreover, as discussed already in the case of energy flows, this generally doesn't represent a grid running on 100% renewables in isolation but rather producing more than the demand in the region. The system is part of a wider grid, which provides power from fossil fuels when renewable energy like wind and solar are not available. It is important that the ultimate goal of a net-zero emissions society is not lost when celebrating some of the renewable energy achievements of these regions. While they represent a significant step forward, it is clear that there is still much work to be done. The pursuit of renewable energy, while essential, is not the only measure needed to mitigate climate change.

3.7 Concluding Remarks

The case studies summarised in this chapter represent great examples of communities coming together to transform their energy system. The goal of all the case study regions, and others that have followed since (e.g. Murau, Austria or Feldheim, Germany), is to become energy independent [3]. To be able to meet all energy needs with local renewable energy instead of having to import fossil fuels. This is an admirable ambition. However, the reality is that it is extremely difficult. The regions have so far failed to get rid of fossil fuels in transport. In the case of Samsø, the vast majority of wind energy is exported, therefore they can say they are net positive on renewables. This hides, for example, the fact that their cars are still driven by fossil fuels. This issue highlights just how difficult it is to eliminate fossil fuels from the energy system, and the massive undertaking that is required across all regions around the world over the coming decades. It also suggests that while the goal of energy independence has been seen to have great impact as a unifying vision for these communities, it may be misguided.

A more impactful ambition would be set in GHG emissions reductions targets as opposed to focusing on energy flows. Following the example of many national governments, community groups could set themselves carbon budget targets. This would better capture all sources of greenhouse gas emissions and ensure more careful planning of mitigation measures. However, it of course brings with it several challenges of its own. There is a higher degree of climate literacy needed to unpack different sources of GHG emissions and understand emissions accounting methods. It is often not possible to get the necessary data for GHG emissions baselines, and thus, common practices rely on what are often crude estimates [61]. This means that there can be wide variations in the emissions determined for an area based on the approach taken to estimate emissions. It is nonetheless clear that more transparent accounting of a region's carbon footprint is important to understand the pathway to a world without fossil fuels.

References

1. P. Droege, 100% renewable: energy autonomy in action, Routledge (2012). https://books.google.com/books?hl=en&lr=&id=2ahMx9eh7m0C&oi=fnd&pg=PP2&dq=100%25+renewable:+energy+autonomy+in+action&ots=Xs4tqebRPo&sig=DemNIXDHTiJj-Z4SoN6Ku01GOZM. Accessed 30 April 2024
2. J.K. Juntunen, M. Martiskainen, Improving understanding of energy autonomy: a systematic review. Renew. Sustain. Energy Rev. **141**, 110797 (2021)
3. M.O. Müller, A. Stämpfli, U. Dold, T. Hammer, Energy autarky: A conceptual framework for sustainable regional development. Energy Policy **39**, 5800–5810 (2011)
4. M. Van Staden, F. Musco (eds.), *Local Governments and Climate Change: Sustainable Energy Planning and Implementation in Small and Medium*

Sized Communities (Springer Netherlands, Dordrecht, 2010). https://doi.org/10.1007/978-1-4020-9531-3
5. J.-A. Perrin, C. Bouisset, Emerging local public action in renewable energy production. Discussion of the territorial dimension of the energy transition based on the cases of four intermunicipal cooperation entities in France. Energy Policy **168**, 113143 (2022)
6. J. Mattes, A. Huber, J. Koehrsen, Energy transitions in small-scale regions–What we can learn from a regional innovation systems perspective. Energy Policy **78**, 255–264 (2015)
7. S. Selvakkumaran, E.O. Ahlgren, Understanding the local energy transitions process: a systematic review. Int. J. Sustain. Energy Plan. Manag. **14**, 57–78 (2017)
8. M. Hecher, U. Vilsmaier, R. Akhavan, C.R. Binder, An integrative analysis of energy transitions in energy regions: a case study of ökoEnergieland in Austria. Ecol. Econ. **121**, 40–53 (2016)
9. P. Devine-Wright, Beyond NIMBYism: towards an integrated framework for understanding public perceptions of wind energy. Wind Energy **8**, 125–139 (2005). https://doi.org/10.1002/we.124
10. C.R. Warren, C. Lumsden, S. O'Dowd, R.V. Birnie, 'Green on green': public perceptions of wind power in Scotland and Ireland. J. Environ. Plan. Manag. **48**, 853–875 (2005)
11. A. Dimitropoulos, A. Kontoleon, Assessing the determinants of local acceptability of wind-farm investment: a choice experiment in the Greek Aegean Islands. Energy Policy **37**, 1842–1854 (2009)
12. M.J. Pasqualetti, Opposing wind energy landscapes: a search for common cause. In: *The New Geographies of Energy* (Routledge, 2013), pp. 218–228
13. N. Hall, P. Ashworth, P. Devine-Wright, Societal acceptance of wind farms: analysis of four common themes across Australian case studies. Energy Policy **58**, 200–208 (2013)
14. M. Wolsink, Planning of renewables schemes: deliberative and fair decision-making on landscape issues instead of reproachful accusations of non-cooperation. Energy Policy **35**, 2692–2704 (2007)
15. J.K. Knudsen, L.C. Wold, Ø. Aas, J.J.K. Haug, S. Batel, P. Devine-Wright, M. Qvenild, G.B. Jacobsen, Local perceptions of opportunities for engagement and procedural justice in electricity transmission grid projects in Norway and the UK. Land Use Policy **48**, 299–308 (2015)
16. K. Schumacher, F. Krones, R. McKenna, F. Schultmann, Public acceptance of renewable energies and energy autonomy: a comparative study in the

French, German and Swiss Upper Rhine region. Energy Policy **126**, 315–332 (2019)
17. P. Vuichard, A. Stauch, R. Wüstenhagen, Keep it local and low-key: Social acceptance of alpine solar power projects. Renew. Sustain. Energy Rev. **138**, 110516 (2020)
18. M. Wolsink, Co-production in distributed generation: renewable energy and creating space for fitting infrastructure within landscapes. Landsc. Res. **43**, 542–561 (2018)
19. P. Späth, H. Rohracher, 'Energy regions': The transformative power of regional discourses on socio-technical futures. Res. Policy **39**, 449–458 (2010)
20. L. Mundaca, H. Busch, S. Schwer, 'Successful'low-carbon energy transitions at the community level? An energy justice perspective. Appl. Energy **218**, 292–303 (2018)
21. A. Huovila, H. Siikavirta, C.A. Rozado, J. Rökman, P. Tuominen, S. Paiho, Å. Hedman, P. Ylén, Carbon-neutral cities: critical review of theory and practice. J. Clean. Prod. **341**, 130912 (2022)
22. K.C. Seto, G. Churkina, A. Hsu, M. Keller, P.W.G. Newman, B. Qin, A. Ramaswami, From low- to net-zero carbon cities: the next global agenda. Annu. Rev. Environ. Resour. **46**, 377–415 (2021). https://doi.org/10.1146/annurev-environ-050120-113117
23. E. Croci, B. Lucchitta, G. Janssens-Maenhout, S. Martelli, T. Molteni, Urban CO2 mitigation strategies under the covenant of mayors: an assessment of 124 European cities. J. Clean. Prod. **169**, 161–177 (2017)
24. C40, 700+ cities in 53 countries now committed to halve emissions by 2030 and reach net zero by 2050. Press Release, April 16, 2021 (2021). https://www.c40.org/news/cities-committed-race-to-zero/. Accessed 3 June 2025
25. R. Koch, M. Hotwagner, Model region Güssing—self-sufficient energy supply based on regionally available renewable resources and sustainable regional development (Austrian Federal Ministry for Transport, Innovation and Technology, 2007)
26. R. Douthwaite, From recession to renewables; Austrian municipality thrives as the world's leading sustainable energy community, Feasta Found. Econ. Sustain. (2006)
27. D. Marcelja, Self-sufficient community: vision or reality? Creating a regional renewable energy supply network, in *Local Governments and Climate Change: Sustainable Energy Planning and Implementation in Small and Medium Sized Communities*. ed. by M. Van Staden, F. Musco

(Springer Netherlands, Dordrecht, 2010). https://doi.org/10.1007/978-1-4020-9531-3

28. H. Hofbauer, Energy from biomass via gasification in Güssing. In: Transform Effect Use Biomass Power World (Pan Stanford Publishing, 2015), pp. 343–373. https://books.google.com/books?hl=en&lr=&id=GJq9BwAAQBAJ&oi=fnd&pg=PA343&dq=Hermann+Hofbauer+gussing&ots=pxvjW8QCwY&sig=M5_2XZxk_jVs_Egi6RD_JKqEUIY. Accessed 15 Aug 2024
29. M. Bolhar-Nordenkampf, R. Rauch, K. Bosch, C. Aichernig, H. Hofbauer, Biomass CHP plant Güssing-Using gasification for power generation, na (2003). https://www.researchgate.net/profile/Markus-Bolhar-Nordenkampf/publication/270160090_Biomass_CHP_Plant_Gussing_-_Using_Gasification_for_Power_Generation/links/54a1863f0cf256bf8baf73e3/Biomass-CHP-Plant-Guessing-Using-Gasification-for-Power-Generation.pdf. Accessed 15 Aug 2024
30. R. Rauch, H. Hofbauer, K. Bosch, I. Siefert, C. Aichernig, K. Voigtlaender, R. Koch, R. Lehner, Steam gasification of biomass at CHP plant Guessing-Status of the demonstration plant, na (2004). https://www.academia.edu/download/73768751/gue_rom.pdf. Accessed 15 Aug 2024
31. JOANNEUM RESEARCH – LIFE: Centre for Climate, Energy & Society, Model Güssing: a vision of energy self-sufficiency, Energy, Climate and air quality, Research and innovation (2015). https://pocacito.eu/sites/default/files/ModelG%C3%BCssing_G%C3%BCssing.pdf
32. Austrian Federal Government, Climate and Energy Fund, Climate and Energy Model Regions An Austrian blueprint for a successful bottom-up approach in the field of climate change and energy; 2018 Update (2018). https://www.klimaundenergiemodellregionen.at/assets/Uploads/Publikationen/2018-Fact-Sheet-Climate-and-Energy-Model-Regions-EN-final.pdf
33. Austrian Federal Government, Climate and Energy Fund, Climate and Energy Model Regions An Austrian blueprint for a successful bottom-up approach in the field of climate change and energy; 2023 Update (2023). https://www.klimaundenergiemodellregionen.at/assets/Uploads/Publikationen/2023-Fact-Sheet-Climate-and-Energy-Model-Regions-EN-final.pdf
34. R.M. De Waal, S. Stremke, Energy transition: missed opportunities and emerging challenges for landscape planning and designing. Sustainability **6**, 4386–4415 (2014)

35. P.J. Jørgensen, S. Hermansen, A. Johnsen, J.P. Nielsen, jan Jantzen, M. Lundén, Samsø-a Renewable Energy Island; 10 years of Development and Evaluation, PlanEnergi and Samsø Energiakademi (2007)
36. A. Radzi, 100% renewable champions: international case studies. In: 100 Cent Renew (Routledge, 2009), pp. 93–165. https://www.taylorfrancis.com/chapters/edit/https://doi.org/10.4324/9781849774345-6/100-renewable-champions-anis-radzi. Accessed 7 May 2024
37. E. Kolbert, The island in the wind. New Yorker **7**, 68–77 (2008)
38. B.V. Mathiesen, K. Hansen, I. Ridjan, H. Lund, S. Nielsen, Samsø Energy Vision 2030; Converting Samsø to 100% Renewable Energy, Department of Development and Planning (Aalborg University, 2015)
39. The World Bank, CO2 emissions (metric tons per capita) - European Union, (2023). https://data.worldbank.org/indicator/EN.ATM.CO2E.PC?locations=EU. Accessed 20 March 2024
40. Smart Rural 27, Smart Rural 27 - Lighthouse example; Samsø (2023). https://www.smartrural27.eu/wp-content/uploads/2023/04/Samso_Lighthouse-factseheet.pdf
41. Samsø Energy Academy, A Guide for Local Pioneer Communities (2016). http://pioneerguide.com/. Accessed 21 April 2024
42. P. Schmuck, S. Eigner-Thiel, M. Karpenstein-Machan, B. Sauer, H. Ruppert, W. Girschner, F. Roland, Bioenergy Villages in Germany: Applying the Göttingen Approach of Sustainability Science to Promote Sustainable Bioenergy Projects, in *Sustain. Bioenergy Prod. - Integr. Approach.* ed. by H. Ruppert, M. Kappas, J. Ibendorf (Springer Netherlands, Dordrecht, 2013), pp.37–71. https://doi.org/10.1007/978-94-007-6642-6_2
43. B. Brohmann, U. Fritsche, K. Hünecke, Case Study: The Bioenergy Village Jühnde, Rep. Öko-Inst. EV Create Accept. Work Package 2 (2006). https://citeseerx.ist.psu.edu/document?repid=rep1&type=pdf&doi=379123583fa264bc7b2363b43b963d29fabc4907. Accessed 7 May 2024
44. H. Ruppert, W. Swantje Eigner-Thiel, F.R. Marianne Karpenstein-Machan, B.S. Volker Ruwisch, P. Schmuck, Paths to the Bioenergy Village Guide, Agency for Renewable Resources (FNR) (2008). https://bioenergiedorf.fnr.de/wege-zum-bioenergiedorf
45. C. Ahl, S. Eigner-Thiel, W. Girschner, T. Göymen, F. Roland, B. Sauer, K. Scheffer, The Bioenergy Village; Self-sufficient Heating and Electricity Supply Using Biomass, Interdisciplinary Centre for Sustainable Development at the University of Göttingen (IZNE)

(2006). http://www.bioenergiedorf.info/fileadmin/user_upload/05-07-04_FaBl_Thesis_20and_20Opportunities.pdf
46. H. Lohrengel, Jühnde Bio-Energy Village in Germany; Civil society engaging in local renewable energy production 2013-11-15 (2013). https://www.eesc.europa.eu/sites/default/files/resources/docs/un_climate_conference_2013_11_1lohrengel.pdf
47. Agency for Renewable Resources, Germany, Home>Bioenergy villages>Overview of bioenergy villages and bioenergy communities, (2024). https://bioenergiedorf.fnr.de/bioenergiedoerfer/uebersicht-der-bioenergiedoerfer
48. Germany's Environmental Protection Agency, Renewable energies continue to pick up speed in 2023 (2024). https://www.umweltbundesamt.de/en/topics/renewable-energies-continue-to-pick-up-speed-in#:~:text=The%20total%20share%20of%20renewable,cent%20in%20Germany%20in%202023
49. A.S.-L. Hauser, A comparative approach to local organisation of the energy transition, PhD Thesis, Universität Würzburg (2020). https://opus.bibliothek.uni-wuerzburg.de/frontdoor/index/index/docId/20210. Accessed 16 Aug 2024
50. L. Watts, Energy at the end of the world: An Orkney Islands saga, MIT Press (2019). https://books.google.com/books?hl=en&lr=&id=Ao-BDwAAQBAJ&oi=fnd&pg=PP8&dq=laura+watts&ots=TlIfQyg1OQ&sig=eGriyWn3CXNvkBqLxXigHwiP8Wo. Accessed 19 Aug 2024
51. Orkeny Islands Council, Orkney Economic Review 2020 (2023). https://www.orkney.gov.uk/media/argiviw0/orkney-economic-review-2020.pdf
52. Northern & Western Isles Energy Efficiency Advice Centre, Orkney Energy Audit; Energy production and use in Orkney 2003-04 (2005). https://www.oref.co.uk/wp-content/uploads/2015/05/Orkney-Energy-Audit-2004.pdf
53. Aquatera Ltd., The Orkney-Wide Energy Audit 2014; Energy Sources and Uses. Report to Orkney Renewable Energy Forum & Community Energy Scotland (2015). https://www.oref.co.uk/wp-content/uploads/2015/05/Orkney-wide-energy-audit-2014-Energy-Sources-and-Uses.pdf
54. Orkney Renewable Energy Forum, Orkney Renewable Energy Forum Aims and Objectives (2024). https://www.oref.co.uk/about-us/aims-and-objectives/. Accessed 19 Aug 2024
55. Aquatera Ltd & Community Energy Scotland, Orkney Energy Audit 2019, 2023

56. K. Johnson, S. Kerr, J. Side, Accommodating wave and tidal energy–Control and decision in Scotland. Ocean Coast. Manag. **65**, 26–33 (2012)
57. M.A. Almoghayer, D.K. Woolf, S. Kerr, G. Davies, Integration of tidal energy into an island energy system—a case study of Orkney islands. Energy **242**, 122547 (2022)
58. Orkney Islands Council, Orkney Hydrogen Strategy; The Hydrogen Islands 2019–2025 (2020). https://www.oref.co.uk/wp-content/uploads/2020/11/Hydrogen-Strategy.pdf
59. G. Dóci, E. Vasileiadou, "Let's do it ourselves" Individual motivations for investing in renewables at community level. Renew. Sustain. Energy Rev. **49**, 41–50 (2015)
60. H.M. Marczinkowski, P. Alberg Østergaard, S. Roth Djørup, Transitioning island energy systems—local conditions, development phases, and renewable energy integration. Energies **12**, 3484 (2019)
61. C. McGookin, B.Ó. Gallachóir, E. Byrne, An innovative approach for estimating energy demand and supply to inform local energy transitions. Energy **229**, 120731 (2021)

4

Irish Environmental Narratives, Policy and Energy Infrastructure

Evan Boyle

4.1 A Brief Note on Historical Foundations of Environmental Narratives in Ireland

Traditionally, environmental NGOs in Ireland campaigned as part of environmental movements or protests. Alongside this, the environmental movement in Ireland is shaped by three historical features [1, 2]. Firstly, it formed around particular issues such as anti-nuking in Carnsore Point, anti-mining in Donegal and anti-pharmaceutical pollution in Cork [3]. On from this, the predominant focus of movements and protests was on the actions of foreign corporations and businesses, premised on the rise of foreign direct investment into the country during this period. Finally, the actors involved were usually local or community-based entities, with additional support from national branches of environmental NGOs. Due to these three factors, localism has been suggested as a defining feature of twentieth-century Environmental activism in Ireland [2].

Throughout the 1990s the state became more open to engaging with environmental narratives, moving away from a previous position that sought to marginalise environmental concern [4]. Following on from

C. McGookin and E. Boyle, *Engaging Communities in Climate Action*, Energy, Climate and the Environment, https://doi.org/10.1007/978-3-031-93498-8_4

the Rio Convention (1992), some organisations in Ireland began to engage with themes of sustainable development. Throughout the 1990s we continued to see protest movements such as the anti-roads protest in the Glen of the Downs, the anti-pylon protests in Cork harbour, protests against Monsanto with relation to genetic engineering, and Cork Harbour Alliance for a Safe Environment (CHASE). Moving beyond the explicitly local characterisation offered, environmentalism in Ireland has been put forward as a dual process; one relating to formal organisations, or official environmentalism, and the other related to local community-based activism [5]. This, however, still forms a limited frame through which to situate environmentalism in Ireland as it moved into the twenty-first century, with the country's position as a globalised state becoming solidified. Environmentalism moved into a space of greater influence with relation to institutional forces, in a cooperative process of acquiescence between different actors involved be they state, NGOs, or community organisations. An institutionalisation of environmental movements can be seen, to some extent, in the Irish context [4] with implications for the development of environmental narratives within Irish society into the twenty-first century.

4.2 A Brave New World of Irish Environmentalism: 2006–7

2006–7 seemed to announce the arrival of a new era in Irish environmentalism. While many of the environmental NGOs that had previously formed part of a protest movement turned their attention to questions of sustainability, the wider social context appeared to become more hospitable to their message. Internationally, Al Gore's film *'An Inconvenient Truth'* (2006) began to cultivate a socio-cultural milieu attuned to the challenge of addressing climate change. The central argument of the film was that climate change can no longer be deemed a political issue. The urgency of action needed means it must be depoliticised, with it instead being framed as the biggest moral challenge facing civilisation at a global level, a proposition currently challenged through Extinction Rebellion. The narrative positioned by the film led to public discourse on the issue, voluntary purchasing of carbon offsets [6], raised awareness and increased concern [7]. Domestically, in 2006, the Irish government launched the Power of One campaign [8, 9] an initiative aiming to encourage consumers to develop more energy-efficient practices. The advertising campaign focused on radio and television adverts, alongside print material delivered door to door, taking an individualist, consumer-orientated approach. The total cost of the campaign was € 10 million. Coming less than two weeks after the Irish release of *An Inconvenient Truth*, the launch of the Power of One campaign aligned with the films call for action at the individual level. 2007 saw work begin on the development of Ireland's first 'eco-village' in Cloughjordan Co. Tipperary, established as a centre for demonstration and learning for sustainable, low-carbon living.

The Stern Review, published in 2006, had brought public attention to the economic imperative of prompt climate policy, and on from this the cost implications that delayed implementation of climate action measures would lead to [10]. The document was not without criticism, and its legitimacy was questioned based on the low cost assigned to emission reduction measures [11]. Alongside this, the Intergovernmental Panel on Climate Change (IPCC) assured the political community of its scientific veracity publishing the first part of its Fourth Assessment Report (AR4), *The Physical Science Basis*, stating that warming of the climate system is unequivocal. This implicit statement was then referenced within Ireland's National Climate Change Strategy 2007–2012, moving policy discussion beyond ambiguity concerning the pressing need for action. Within the strategy, considering the ratification of the Kyoto Agreement in 2005, the government outlined plans for the period 2008–2012, where a reduction of 12% from the current level (2006–7) would be achieved.

In June 2007, following a general election, Fianna Fáil and the Green Party ratified a Programme for Government. The Green Party's 25-year adolescence in grassroots politics was over, with it now taking its place in the politics of government. The organisational transformation of the Green Party towards 'professional' politics was something not unique to Ireland. Tracing the development and progression of Green parties across a number of both European and non-European contexts, Frankland et al. [12] highlight a common trajectory through which parties founded in environmentalism and grassroots protests mature and are institutionalised into the established political system, in some cases forming coalitions with conventional parties of power. In the Irish case, Green delegates voted 441 to 67 in favour of joining a coalition with Fianna Fáil and the Progressive Democrats. Two senior ministerial positions were given to the Green Party in important areas aligned with the Green Party agenda; the Department of Environment, Heritage and Local Government and the Department of Communications, Energy and Natural Resources. This positive outcome for the Greens was not without cost. In joining the government with 6 seats, the party played an important role in government but was not indispensable. Compromises were needed on different value-orientated issues. While the emergence of the Green Party from community protestors to minor collation partners is "one of the more interesting political events in the history of the nation's environmental movement" it contained sacrifice with the party's "association with rural community groups (coming) to be diminished as part of their pursuit of power" [13, p. 68]. A tension emerged between the party members in government and the external party members. The proposal for a new motorway through the Hill of Tara and the continued use of Shannon Airport for American troops were problematic issues within the party [14]. Perhaps most contentious, however, was Minister of Communications, Energy and Natural Resources, Eamon Ryan, overseeing the Corrib Gas Project when in office, a development which he staunchly protested against before the formation of the 30th Dáil. This contradiction can be referenced as an example of the restrictions placed on the individual within the institution of office, or more controversially, the coercive and hegemonic social, economic and political organisation of Irish society [15].

4.3 Targeting Transition and Taming the Tiger: 2008–2012 (1268)

In early 2007, the European Council approved the objective of reducing emissions of greenhouse gases, brought forward by the European Commission. As part of this, the 20–20-20 targets were established whereby nations were asked to commit nationally to a 20% reduction. In 2008, alongside the other 26 EU Heads of State, Taoiseach Brian Cowen agreed to the mandated target of a 20% reduction in greenhouse gas emissions by 2020 compared to 1990 levels. Also, that year, the first referendum on the Lisbon treaty was held. Discussion on the environmental aspects of the treaty brought about an increase in media discussion on climate change in Ireland [16].

The graph below (Fig. 4.1) highlights the trends in climate change coverage from 1997 to 2011. The Irish Times is the newspaper which gives most significance to climate change over this period. The authors relate the trend to Downs' [17] "Issue Attention Cycle". The basis of this theory on public attention to domestic policy was one of a circular nature. In the Irish case, the slow development of climate change coverage over the period 1997 to 2005–6 is evident. The issue is placed in the early and pre-problem phase. A sudden and sharp increase is apparent in 2006–7. This increase can be correlated to some degree with the number of factors referenced above from domestic to international, policy to media. Following this, "*coverage then plateaued above two per cent for two years during the realisation of costs stage, before falling dramatically during the post problem stage*" [16, p. 22].

While the authors suggest that Downs' theory of "Issue Attention Cycle" may be relevant they are also clear in outlining the impact which the global financial crisis of 2008–9 may have had on climate change coverage in the mainstream media. Ireland was one of the worst impacted European countries during the global recession and "*in a world where newspapers have to focus on the here and now, it is possible that the immediacy of one crisis could push aside the gravity of a seemingly less pressing one*" [16, p. 22]. The decreased reporting in the media of environmental issues correlates to a decrease in public consideration of the issue as a key challenge facing the country. The graph below (Fig. 4.2), illustrates

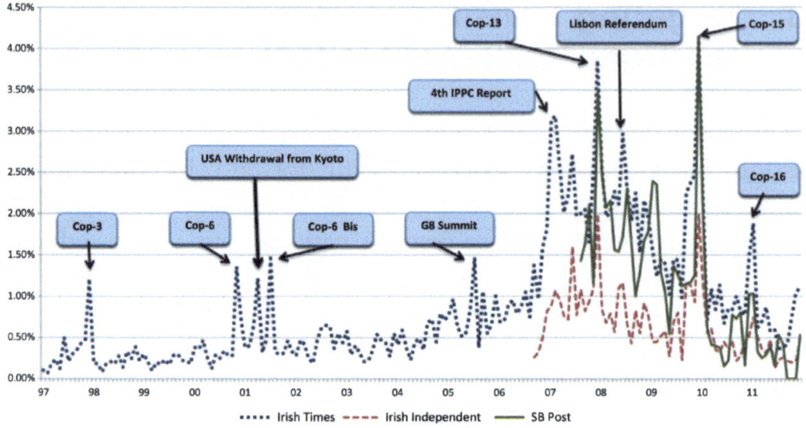

Fig. 4.1 Newspaper trends in climate change coverage [16]

the fall in public perception of the environment, climate and energy as a central concern for the citizenry considering the economic crash. Ireland represents a more pointed example of a European wide trend, whereby as of 2009 the environment, climate and energy are positioned at the lowest position of public importance over the decade from 2006 to 2016.

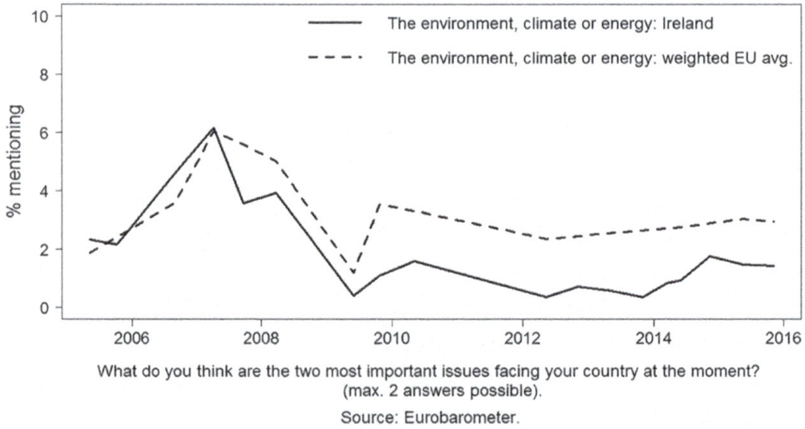

Fig. 4.2 Public perception of the environment, climate and energy in Ireland and the EU 2006–2016 [18]

In the years following the 2008 economic crisis, the Irish Government undertook a €64 billion bank bail-out with the IMF and the EU leading to increased austerity measures within policy [18, 19]. Alongside this, unemployment rose by nearly 8% between 2008 and 2010. The recession brought with it a significant reduction in carbon emissions, down from roughly 67 million tonnes per annum between 2005 and 2009 to 57 million tonnes in 2011. This trend seems strongly related to economic factors, rather than any sense of a new environmental consciousness solidifying in Irish society. There were further sustainability triumphs following the collapse of the Celtic Tiger, with more sustainable consumption practices evident in the country [20]. Again, this is systematic of austerity and unemployment as opposed to an intensified environmental narrative. By 2010–11 discussions on environmentalism significantly decreased within the country's dominant media sources [16]. Considering the global economic crisis, the 2009 conference of the parties (COP) in Copenhagen proved unsuccessful in building and improving upon the ambition set out through the Kyoto protocol and the 20–20-20 plans. The targets set out in Copenhagen were received unenthusiastically within the climate science community, with doubts suggested about their ability to keep below even 3 °C [21]. Across Europe, in light of the Copenhagen Accord, there was a "*reorientation of EU priorities, driven significantly by the economic and fiscal crisis*" *with "a noticeable dampening of the enthusiasm of the 2000s for climate policy in the EU*" [22, pp. 250–251].

As Ó Riain illustrates, the most prominent feature of Ireland's economic crash is "the steady erosion of public trust and political legitimacy, even in the absence of widespread disruptive protest" [23, p. 259]. Despite little civil society unrest during the initial stages of the banking crisis, the 2011 general election brought about an 'earthquake election' [23]. Following the passing of the January Budget, the Green Party stepped down as a minor coalition partner leading to the election. Fianna Fáil gained only 17% of the votes, losing 58 seats from the 2007 election in what was described as "the most volatile (election) in Irish history and the third most volatile in the history of Western Europe since 1945" [24, p. 1308]. Alongside this, the Green Party lost all their seats [23]. While in government, the economic collapse had "*delayed the roll-out of many of the environmental investment programmes promised by the Green Party*" [25, p. 348]. They did, however, manage to introduce several successful policy measures in this time such as a reform of planning regulations, funding for energy efficiency retrofitting projects, and measures to mitigate against GHG emissions [24]. The Green Party, as a junior coalition partner, suffered significantly at the ballot in 2011, aligning with the fears presented by some party members in 2007 who did not see Fianna Fáil as "the ideologically most preferred coalition partner" [14, p. 615]. The "fear of contamination" was realised [24]. At this point, the media turned its attention to attacking the Green Party [25].

Civil society unrest with regard to the environment manifested around fracking, despite the lack of widespread unrest in a political sense. Licences for the shale gas development through hydraulic fracturing were initially introduced by the Irish Government before a 2011 moratorium was passed requiring the completion of a 2013–2016 best practise report [26]. The central governmental concern towards managing a way out of austerity led to proposals for economic growth, independent of environmental impact [15]. While the best practice report was being drafted a significant rural mobilisation against potential developments surfaced. This social movement followed on from the campaign around the Corrib gas field; "*the long-standing struggle over gas drilling in northwest Ireland has been extended by fracking proposals covering much of the West Midlands*" [27, p. 5]. National media reporting on the topic brought the anti-fracking campaign into the public consciousness. The framing of activists was centred on the approach taken, the ability to reason and the degree of localism. All these factors were ultimately linked to their position on economic development [26]. The application of pressure on the government to not go ahead with fracking proposals, mobilised in many cases at the local level, may have played a role in the ultimate decision to ban onshore fracking in the country [27]. The communitarian power of protest in Ireland [28] was again shown through the Right 2 Water campaign of 2014–2016 [29].

This period also saw experiments with deliberative democracy in Ireland, with an importance placed on awareness-raising, capacity building, community participation and partnership models. In 2011, the experimental pilot study known as "We the Citizens" was developed to investigate citizens assembly approaches to deliberative democracy in the Irish context. This development took place to strengthen democratic processes considering negative citizen sentiment towards political processes in light of austerity.

4.4 Recovery—Reframing Sustainable Development: 2012–2016

In 2012 the National Economic and Social Council (NESC) published a report on climate change in which it was stated that the issue must become ubiquitous across policy, engaging "*actors at all levels and in all sectors, through a governance system that animates, learns from and pushes networks of firms, public organisations and communities to ever-greater decarbonisation*" [30, p. 11]. The publication of the report led to "*the most informed public debate so far*" about decarbonisation [31]. Over several weeks, articles were published within *The Irish Times* discussing different aspects of the report, such as a lack of reduction targets and the agriculture sector. What this public debate failed to outline, Kirby [31] notes, was the "positive thrust" of the report, with the issue of emission reduction and climate change being put forward as a central policy concern for the Irish government moving forward. In 2014 the economy began to stabilise, with GDP steadily increasing from a 2013 low point and unemployment rates continuing on a downward trajectory. Emissions were steadily rising again after the decline from 2008 to 2011 "*indicating that Ireland's greenhouse gas emissions remain tightly coupled with economic growth*" [32, p. 9]. Around this time the Climate Action and Low Carbon Development Act mandated the creation of a National Mitigation Plan. This was later published in July 2017, a full decade from the last national climate change plan [30].

Two international developments coincided with the mandate for a new climate action plan. Firstly, the concept of the Sustainable Development Goals (SDGs) was implemented. A global sustainability panel set up as part of the Rio + 20 summit in 2012 made recommendations for the creation of Sustainable Development Goals (SDGs) to continue from the Millennium Development Goals (MDGs). It was heralded as "*an important idea and could help finally to move the world to a sustainable trajectory*" [33, p. 2206]. Across 17 sectors the SDGs are orientated towards sustainability throughout all aspects of society, creating universally held standards that all nations adhere to. Optimistically it was claimed that "*the goals may help create a fresh global narrative around a common future to mobilise collective action and help develop a shared understanding of interconnected risks and solutions*" [34, p. 23]. This aligns with the calls for collective conscience on the back of the failure of COP 15 as shown by Deane-Drummond [34]. Yet, the relative lack of interest in the SDGs in the public and the media was noted [35], with them failing to play a role in influencing the narrative due to cognitive links to the MDGs and issues of development in countries of a lower economic profile. A framing bias influenced both public and political understanding of the SDGs, as they were unsuccessful in "*form(ing) part of the national discourse… the public do not typically view progress and development in those terms, and consequently many governments may not either*" [35].

Secondly, the 21st Conference of the Parties (COP21) was held in Paris in 2015. The first major COP since the abject failure of COP15 and the global economic crisis, the Paris Accord succeeded in bringing together 195 countries, a diplomatic success with relation to climate action. Despite the obvious failure of COP15, it was a significant development in international climate governance by creating a mechanism through which all states could contribute to collective mitigation; "*a collective, 'bottom up' approach in which all nations voluntarily pledged their emission-reduction targets*" [36]. This was the groundwork for the emergence of nationally determined contributions (NDCs) within COP21. The NDCs act as detailed plans to reduce GHGs at a domestic level [37]. Reporting on COP21 appeared across the Irish media, representing an established order whereby international climate-related events gain more coverage than national or local climate-related news [38]. A danger exists whereby through this disposition towards international coverage, an environmental narrative may appear around the imposition of climate policy and climate action from foreign entities. While this sentiment exists, it is currently peripheral within Irish society.

As the climate change debate moved to the post-scientific consensus stage [39], a moral dilemma took hold for institutions that operate to benefit the public, such as charities and public institutions. This called into question investments that seek to profit from the continued use of fossil fuels in a time when the environmental narrative had become dominated by the irrefutable need to transition to a low-carbon society. As policy moves towards sustainability, in keeping with the sustainable development goals (SDGs) and the agreements of COP21, these investments also represent potentially stranded assets [40]. The Fossil Fuel Divestment (FFD) movement emerged in this space as an international mobilisation on moral grounds [41, 42], perhaps as a foreshadowing of morally focused protest forms which have arisen in more recent times. The progress of the Irish FFD movement has been one of the most successful instances. The divestment campaign in Ireland was led by civil society organisations such as Trócaire, An Taisce and Good Energies Alliance Ireland along with university campaigns in NUI Maynooth and Trinity College Dublin. It sought for the divestment of €72 million which the government has invested in non-renewables through the Irish Strategic Investment Fund (ISIF). Third sector mobilisations and campaigning acted to exert pressure upon the government, resulting in a divestment bill being brought to parliament; "*Trócaire and the others…have fought the battle and forced us into this situation in the Dáil*" [41]. The idea of energy citizenship began to materialise in Irish policy and discourse in 2015, with environmental groups, energy groups and civil society organisations all becoming involved in the conversation [42].

4.5 Deliberative Democracy and Climate Policy: 2016–2019

The Citizens' Assembly on Climate Change was established by Fine Gael as part of a commitment made during the 2016 election campaign. The assembly "*followed closely the model of the 2012–14 Irish Constitutional Convention in having as its members 99 citizens selected at random*" [43, p. 114], to convene over two weekends to deliberate on information regarding different issues, and built upon previous experimentations with deliberative democracy in the Irish context such as the constitutional convention [43], the We the Citizens pilot citizens' assembly [44], and processes emanating from Local Agenda 21. Five topics were selected for deliberation: abortion, the challenges and opportunities of an ageing population, fixed-term parliaments, how referenda are held, and how the state can make Ireland a leader in tackling climate change. The Citizens Assembly on Climate Change was held between October and November 2017. Under intense international pressure for action on climate change due to the Paris Accord and Ireland's miserable failure to get close to the 2020 emissions targets [45], the climate was selected as a suitable topic of deliberation through the citizens' assembly model [46]. There were also 1,185 public submissions made to the process. From these two processes, 13 recommendations were made followed by the creation of a special parliamentary committee to consider the finalised report [46].

The deliberations reiterated the centrality of citizens and communities [47]. The recommendations made by the assembly were far more progressive than any proceeding government plan or framework, including an 80% willingness to pay a carbon tax and 100% of the Members recommending that the Government should take a leadership role in addressing climate change through mitigation measures. Of the 13 recommendations drafted by the Citizens Assembly on Climate Change, one recommendation made explicit the need to enable a feed-in tariff for micro generation and the need to stimulate community ownership of renewable energy projects moving forward. In June 2018, as part of the Project Ireland 2040 initiative, the government held a discussion entitled "Empowering Communities for Climate Action".

Throughout this period SEAI began encouraging communities to sign up to the Sustainable Energy Community (SEC) network, through which mentoring support would be provided to groups undertaking energy initiatives. Close to 800 groups had signed up to the SEC network as of December 2023. The Department of Communications, Climate Action and Environment's publication of the National Mitigation Plan on Climate Change highlighted the importance of community engagement, participation and acceptance with regard to decarbonising electricity generation. Throughout the preceding twenty years the institutional inclusion of 'community' within the socio-technical transition to a low-carbon society, although not to suggest its absence, had been incoherent and piecemeal.

A Joint Oireachtas Committee on Climate Action (JOCCA) was set up in 2018 to review the recommendations of the Citizens' Assembly. The report from the JOCCA "*potentially represents a significant departure in Ireland's response to climate change*" [32, p. 11], with it, and the citizens' assembly, coming from domestic concerns rather than international pressure. Torney and O'Gorman show a growing public awareness emerging in Ireland with relation to the necessity for urgent climate action. The JOCCA report also represents a sliding doors moment for the media, specifying the need for greater coverage and content on climate change, alongside dedicated programming [48]. While Ireland does not suffer from "false balance" [49] to the same extent as other countries, increased exposure of an environmental narrative within the main broadcaster would not be unwarranted. Weather events such as storm Ophelia ('the beast from the east') and the summer drought and fodder crisis were all referenced in the media with relation to climate change impacts [50]. The 2019 SFI Science Week, a nationwide initiative for science education, focused specifically on climate action with all major media outlets in the country focusing on the issue; not in the sense of debating the science of climate, but rather debating what actions that should be taken considering the science. The regularity with which the need to transition became referenced in the Irish media in 2019 was unprecedented.

From policy to civil society movements, 2019 was filled with discourse on climate change, and with this, the necessity of climate action was widely recognised. Starting with civil society movements we can look firstly to the Fridays for Future youth climate protests [51], which appeared in Ireland—and internationally—in March of 2019, with participation found across the country. The movement continued throughout the year, seeing both large- and small-scale mobilisations across Ireland, culminating in the 20th September 2019 climate strikes. Alongside Friday's for Future, Extinction Rebellion was established as a mechanism of protest through coordinated non-violent civil disobedience [52]. A wide range of protest methods were deployed by the movement, including an instance where protesters chained themselves to Dáil Eireann and covered an external wall in exerts from scientific journals. Alongside this, an encampment of protestors formed outside Leinster House in swimwear, flip-flops and sunglasses on a rainy day in October to demonstrate against comments made by the Taoiseach suggesting that Climate Change could bring some benefits to Ireland such as warmer weather.

This is not to suggest that the environmental narrative in Irish policy has been one of dismissal. The publication of the Climate Action Plan (2019) illustrated cross-governmental plans to implement the changes required to adopt a more ambitious commitment of net-zero greenhouse gas emissions by 2050 [53]. Although lacking in some regards (e.g. continued dependency on private cars, omission of native woodlands for carbon capture and storage), the 183 actions presented offered considerable optimism for Ireland to move beyond its current position as a climate laggard [45]. Prior to the publication of the Climate Action Plan, Ireland became only the second country in the world to declare a climate and biodiversity emergency. The declaration was later halted by the Minister for the Department of Communications, Climate Action and the Environment, Richard Bruton, as it was deemed to not align with the proposals set forth for decarbonisation within the Climate Action Plan.

At the UN Climate Summit in September 2019, the Taoiseach announced the Governments plans to phase out fossil fuel exploration, despite this not featuring within the Climate Action Plan. While governmental commitment continued to appear incoherent, widespread societal discourse on the topic had emerged, placing it at the forefront of political discussion. This was illustrated through the 2019 local elections and the success of the Green Party in gaining seats. From civil society movements to national politics, the final year of the decade saw the blossoming of environmental narratives as central to public discourse. The central importance of climate action became embedded in any discussion of Ireland's development in 2019.

4.6 COVID-19 and the Green Deal: 2020–2021

With 2020 and the first months of 2021 so firmly defined by the COVID pandemic, environmental narratives took something of a back seat. There was, however, some cause for optimism with regard to aligning these two societal challenges illustrated, for example, by the EU Green Deal. Lessons could be learnt from our crisis response which can be applied to the environmental challenges faced moving forward. The pre-COVID landscape was one in which environmental narratives had gained a primacy that had not been seen in a long time, if ever. During the infancy of the COVID-19 pandemic, with strict lockdowns imposed across the world, narratives that "nature is healing" implying improvements to the environment attributed to human absence gained salience. Lockdown measures also had a clear impact on annual CO_2 emissions globally, with estimations that 438Mt of carbon emission reduction was seen across 184 countries in 2020 when compared with 2019 [54].

Despite the centrality of the pandemic, the environment still held sway in political discussions. The General Election saw the re-emergence of the Green Party in coalition with Fianna Fáil and Fine Gael and a much deliberated upon agreement to cut carbon emissions by 7% annually to meet 2030 targets was implemented in policy. While the feasibility of the ambition must be investigated [55], the centrality of emissions reduction during what was ostensibly a war footing Government must be seen as a strong commitment with regard to reducing fossil fuel use during the decade of the 2020s. The 2020 Programme for Government emphasised the importance of climate governance, emission reductions and building stronger communities. The 2021 Budget announced an increase in the carbon tax, increased support for retrofit programmes and a holistic approach to sustainable transport across walking, cycling and public transport. In October of 2020, the Government published a new carbon law that committed Ireland to net-zero carbon emissions by 2050. A €15.4 million grand package was also announced for digital hubs and remote working facilities in early 2021.

The Renewable Electricity Support Scheme (RESS) granted support to seven community energy projects in Ireland in 2020, a landmark moment within the community energy sector in Ireland. History, however, makes clear the need for healthy scepticism. As has been shown in the past [56] 'bottom-up' community involvement can often be driven by the failings of 'top-down' approaches. In the Irish context, the failings with regard to 2020 targets may have led to new thinking with regard to how to approach the energy transition. While cautiously a change in policy vision towards more participatory and engaged processes can be proposed, this is in no way suggestive of success, with close attention needed to how this plays out in the real world, particularly considering it being a novel method of implementation from a policy perspective, and the changing nature of environmental narratives through a range of factors be they economic, political or external. COVID-19 shifted the focus of discussions on climate action from "can we" to "will we", with this move having an impact with regard to political leadership, scientific evidence, societal buy-in, and market forces.

Over the proceeding decade, national policy around Ireland's low-carbon transition had increasingly focused on public participation, with terms like 'energy citizen' and 'sustainable energy communities' being used. One of the earliest mentions was in the Irish Government's Energy White Paper from 2007, which noted that the renewable energy transition could play an important part in rural/regional development and endorsed greater community involvement. This was reinforced in the Strategy for Renewable Energy 2012–2020, which noted that achieving renewable energy targets must be a "collective endeavour" and highlighted that one of the significant challenges was securing local benefit and winning public acceptance [57] The Programme for Government agreed in June 2020 had a subsection dedicated to supports for the community energy sector "Bringing Communities with Us" and another outlining "A New Engagement Model", which stated that to "*develop a new model of engagement with citizens, sectors, and regions as an early priority for Government*" (Department of the Taoiseach, 2020). Following this, in March 2021, the Department of the Environment, Climate and Communications launched the Climate Conversations seeking both expert and citizen submissions for the Climate Action Plan 2021 [58].

Internationally, January 2021 marked the rejoining of the Paris agreement by the United States having left following the election of Donald

Trump in 2016. Setting ambitious targets for cutting greenhouse gas emissions by 50–52% from 2005 levels by 2030, President Biden's administration reinstated a commitment to global partnership on decarbonisation in the years ahead. In August of that year, the 6th IPCC assessment report was published, highlighting the inextricable impact of human activity on the climate with warming taking place within the atmosphere, on land and in the ocean. Immediate action was called for to prevent warming beyond 1.5^{O}C. Towards the end of the year, in November, COP26 was held in Glasgow with a renewed focus on limiting global warming through national pledges and increased emphasis on annual climate financing to support developing countries. Also at the global level, 2021 saw extreme weather events from heatwaves in North America, flooding in Europe and Asia, severe droughts in East Africa, and wildfires in the United States and Europe.

4.7 Ukraine, Energy Poverty and Plan-Led Deployment: 2022–2023

On the 22nd of February 2022, the Irish Government announced an end to all COVID restrictions including vaccine certificates, social distancing and household visits. The National Public Health Emergency Team (NPHET) was disbanded. On the 24th of February, the Russian invasion of the Ukraine marked the largest military attack on a European country since the Second World War. The humanitarian crisis of displaced Ukrainians led to the external migration of over eight million people, a refugee crisis the scale of which had not been seen in close to a century. Beyond the primary horror and catastrophe of the invasion, the far-reaching impacts were felt globally. Within the European context, the coupling of the European power system to Russian gas was exposed as a geopolitical miscalculation. The policy response to the invasion and its immediate energy consequences had expedited energy policy and strategy at EU level towards efforts to diversify towards alternatives such as renewable energy solutions, the prioritisation of energy security through investment in new infrastructure including storage and interconnectors and an accelerated approach to the energy transition across member states. Despite this increased urgency, some wider challenges were further exposed as Europe sought to respond to the implications of geopolitical tensions with Russia. In exposing energy as a geopolitical concern, new policies can negatively impact fossil fuel phase out and cause issues in relation to energy equity. A trade-off occurred in some member state contexts whereby sustainability and equity were marginalised in favour of geopolitical security as the driving force of the energy transition. Yet, still the potential for an acceleration of clean energy supply within Europe was obvious [59].

Domestically, the invasion led to a sharp rise in inflation (reaching 10%) and spiralling energy prices. In 2022, around 70,000 Ukrainian refugees fled the war to Ireland. As referenced, while the reception to these migrants was almost universally welcoming and supportive, the sheer number of immigrants coupled with those from other countries began to raise some apprehensions in light of the pressures felt on the housing market. Some protests were seen throughout the country [58]. Despite these instances, a Eurobarometer poll showed public satisfaction with the Governments response to the invasion of Ukraine was the second highest level in Europe. The financial implications remained acute for households with 68% of people surveyed said the war "had serious financial implications for them". Concerning energy, an ERSI report published in June 2022 found that an estimated 29% of households in the Republic qualify as experiencing energy poverty. The coalition government, in both 2022 and 2023, introduced a range of measures and supports to mitigate the impact of rising energy prices for homes and businesses. Alongside a range of specific targets such as increased public transport use and a move towards EVs, the publication of the Climate Action Plan for 2023 [58] outlined targets towards the accelerated delivery of renewables including at least 500 MW of local community-based RE projects. The updated CAP also highlighted citizen engagement as a key aspect of Ireland's transitory vision for climate action. The expected outcome of such citizen engagement strategies is expressed as "*a strengthening social contract between the Government and the Irish people around climate action*" [58].

In 2023, in the landscape of geopolitical tensions on security of supply, Irish offshore energy policy set a clear strategy to develop 5GW of offshore energy by 2030. A government led offshore wind taskforce was established aiming to ensure that the first phase of offshore projects (2–4GW from 2027 to 2030) was given a clear pathway to the planning system through the ORESS auction. Towards the end of 2023 four successful projects were announced, three on the east coast and one on the west coast ranging from 450 MW to 1.3GW. This represented the largest volume of renewable energy ever awarded at auction in Ireland. The move towards a plan led approach to offshore renewable development also sought to make sure the second phase of projects (making up the balance of first 5GW) have a clear route to market and support within the planning system. A final objective of the task force on offshore wind was to publish a policy to harness the overall 30GW moving forward. This strategic change towards a plan led approach to development as opposed to a developer led approach, aimed to provide clarity for the development of the resource.

Also in 2023, first Ireland's Climate Change Assessment (ICCA) report was published led by the Environmental Protection Agency. Overall, across Ireland, average temperatures were shown to be approximately 1 °C higher than the early twentieth century with 16 of the 20 warmest years occurring since 1990 and 2022 being the hottest year on record to date. On a per capita basis, Ireland was shown to be the second highest ranked country for all GHG emissions. A late action scenario has been predicted as leading to temperature increases of 2.77 °C which would lead to increased storms, intense precipitation and further warming. The report goes on to suggest the strategic opportunities which transformative actions will have in relation to mitigation and adaptation and the associated health, wellbeing and sustainable economic benefits of such. Within the ICCA, the state is named as having a central role to play in ensuring the necessary measures are taken for the wellbeing of future generations.

A citizen's assembly on biodiversity was held in Ireland in 2022, with a final report submitted in April 2023. Ninety-nine randomly selected members of the public convened over two weekends before agreeing upon 159 recommendations, including 73 high-level recommendations and 86 sectoral specific actions and priorities. The recommendations of the assembly found that strong leadership, a clear vision and enabling conditions for action were required. It found that action to date in implementing and enforcing biodiversity-related laws and directives has been a comprehensive failure, with a need for increased ambition, clear planning and communication, and changes in the funding, implementation and enforcement of policies. The constitutional protection of nature was also suggested. Following the publication of the citizens' assembly report, the Joint Committee on Environment and Climate Action recommended that the government advance a referendum on protecting biodiversity. Such a move would make Ireland the first country in the EU to enshrine the rights of nature into a national constitution.

4.8 Final Remarks

Across the two decades covered within this chapter we see a clear increase in policy significance and media covered overall. However, within this, the politicised and popularised nature of climate action means the challenge is not isolated from landscape developments be they geopolitical, economic or cultural. Such landscape develops are continuing from the time of writing this chapter, to its publication, and your reading. The local benefits of the energy transition must be central to notions of the transition due to the increasingly unstable nature of global politics. To respond effectively and urgently limit global warming to the best of our collective abilities, it must transcend political dichotomies. However, what we are seeing in 2024 is increased polarisation and the escalation of divisions within the citizenry. In 2024 an unprecedented electoral cycle took place with 76 countries, including major global powers, holding national elections during the year. Democratic stability is in question due to increased populism and geopolitical divides. Global diplomacy for collective climate action is in no way guaranteed in the current political atmosphere. However, it remains key to the agenda and global shifts towards integrated approaches to renewable deployment, sustainability financing, nature-based solutions, and energy infrastructure give credence to optimism in the face of a potential reordering of global power structures.

References

1. S. Yearley, The social shaping of the environmental movement in Ireland. In: *Irish Society: Sociological Perspectives*, D. P., L. S., K., and L. O'Dowd (Eds.) (Dublin: Institute of Public Administration, 1995)
2. M. Garavan, Resisting the costs of "development": local environmental activism in Ireland. Environ. Polit. **16**(5), 844–863 (2007)
3. S. Baker, The evolution of the Irish ecology movement. In W. Rudig (Ed) *Green Politics I'* (1990)
4. G. Mullally, Relocating protest: globalisation and the institutionalisation of organised environmentalism in Ireland? In: *Social Movements in Ireland*, L. Connolly and H. N, Eds. (2006), pp. 145–67
5. R. Allen, *No Global: The People of Ireland Versus the Multinationals* (Pluto Press, London, 2004)
6. G.D. Jacobsen, The Al Gore effect: an inconvenient truth and voluntary carbon offsets. J. Environ. Econ. Manag.Manag. **61**(1), 67–78 (2011)
7. J.M. Nolan, An inconvenient truth" increases knowledge, concern, and willingness to reduce greenhouse gases. Environ. Behav.Behav. **42**(5), 643–658 (2010)
8. S. Diffney, S. Lyons, L. M. Valeri, Advertising to boost energy efficiency: the Power of One campaign and natural gas consumption, vol. 280. ESRI Working Paper (2009)
9. S. Diffney, S. Lyons, L.M. Valeri, Evaluation of the effect of the power of one campaign on natural gas consumption. Energy Policy **62**, 978–988 (2013)
10. N. Stern, *The Economics of Climate Change: The Stern Review* (Cambridge University Press, Cambridge, 2007)
11. R.S. Tol, G.W. Yohe, A review of the Stern review. World Econ. Henly Thames 7(4), 233 (2006)
12. E.G. Frankland, P. Lucardie, B. Rihoux, *Green Parties in Transition: The End of Grass-Roots Democracy?* (Ashgate Publishing Ltd., Surrey, 2008)
13. L. Leonard, S.B. Kedzior (eds.), *Occupy the Earth: Global Environmental Movements* (Emerald Group Publishing, Bingley, 2014)

14. N. Bolleyer, The Irish Green Party: from protest to mainstream party? Ir. Polit. Stud. **25**(4), 603–623 (2010)
15. A. Slevin, *'Consent, Coercion and Consequences of the Corrib Gas Conflict', in Gas, Oil and the Irish State* (Manchester University Press, Manchester, 2016)
16. P. Wagner, D. Payne, Trends, frames and discourse networks: analysing the coverage of climate change in Irish newspapers. Ir. J. Sociol.Sociol. **25**(1), 5–28 (2017)
17. A. Downs, Up and down with ecology: the" issue-attention cycle. Polit. Am. Econ. Policy Mak. **48** (1996)
18. D. Hanley, Austerity Ireland: the failure of Irish capitalism. J. Contemp. Eur. Stud. **23**(2), 297–298 (2015)
19. P. Clancy, T. McDonnell, *From Tiger to Bailout: The Irish Crisis* (ETUI, Brussels, 2011)
20. F. Murphy, Austerity Ireland, the new thrift culture and sustainable consumption. J. Bus. Anthropol. **6**(2), 158–174 (2017)
21. J. Rogelj et al., Copenhagen Accord pledges are paltry. Nature **464**(7292), 1126–1128 (2010)
22. S.Ó. Riain, *The Rise and Fall of Ireland's Celtic Tiger: Liberalism, Boom and Bust* (Cambridge University Press, Cambridge, 2014)
23. M. Gallagher, M. Marsh (eds.), *How Ireland Voted 2011: The Full Story of Ireland's Earthquake Election* (Springer, New York, 2011)
24. P. Stafford, The 2011 general election in Ireland: the political impact of the Irish recession. Representation **47**(3), 343–356 (2011)
25. G. Mullally, Fear and loading in the Anthropocene Narratives of apocalypse and salvation in the Irish media. In: *Transdisciplinary Perspectives on Transitions to Sustainability* (Routledge, Abingdon, 2017)
26. T. Steger, A. Drehobl, The anti-fracking movement in Ireland: perspectives from the media and activists. Environ. Commun.Commun. **12**(3), 344–356 (2018)
27. L. Cox, Challenging Austerity in Ireland. Concept **3**(2) (2012)
28. A. Germanos, Victory declared as Ireland bans fracking. Green Left Wkly. 1143(1), 15 (2017)
29. C. Ritter, Digital Dublin: the water protests and social media in Ireland. AoIR Sel. Pap. Internet Res. **6(1)**, 1–3 (2016)
30. P. Kirby, Policy optimism: NESC, climate change and achieving decarbonisation. Administration **61**(2), 75–90 (2013)

31. D. Torney, R. O'Gorman, A laggard in good times and bad? The limited impact of EU membership on Ireland's climate change and environmental policy (2019)
32. J.D. Sachs, From millennium development goals to sustainable development goals. The Lancet **379**(9832), 2206–2211 (2012)
33. O. Gaffney, Sustainable development goals. SDGs, transforming our world:2030. Glob. Change **82**(1), 20–23 (2014)
34. C. Deane-Drummond, A case for collective conscience: climategate, COP -15 and climate justice. Stud. Christ. Ethics **24**(1), 5–22 (2011)
35. S. MacFeely, Measuring the sustainable development goals : what does it mean for Ireland? Administration **65**(4), 41–71 (2017)
36. P. Christoff, The promissory note: COP 21 and the Paris Climate Agreement. Environ. Polit. **25**(5), 765–787 (2016)
37. A. Vogt-Schilb, S. Hallegatte, Climate policies and nationally determined contributions: reconciling the needed ambition with the political economy. Wiley Interdiscip. Rev. Energy Environ. **6**(6), 256 (2017)
38. E. Fox, H. Rau, Disengaging citizens? Climate change communication and public receptivity. Ir. Polit. Stud. **32**(2), 224–246 (2017)
39. A.J. Hoffman, *How Culture Shapes the Climate Change Debate* (Stanford University Press, California, 2015)
40. J. Ayling, N. Gunningham, Non-state governance and climate policy: the fossil fuel divestment movement. Clim. Policy **17**(2), 131–149 (2017)
41. N. Antadze, K.A. McGowan, Moral entrepreneurship: thinking and acting at the landscape level to foster sustainability transitions. Environ. Innov. Soc. Transit. **25**, 1–13 (2017)
42. C. Watson, E. Boyle, G. Mullally, B. O'Gallachoir, *Responding to the Energy Transition in Ireland: The Experience and Capacity of Communities* (Department of Sociology and School of Engineering, MaREI Centre, Environmental Research Institute, University College Cork, 2020)
43. D.M. Farrell, J. Suiter, C. Harris, Systematizing'constitutional deliberation: the 2016–18 citizens' assembly in Ireland. Ir. Polit. Stud. **34**(1), 113–123 (2019)
44. A. Vergés Bausili, *The Constitutional Convention and Ireland. Ireland and The European Union: Nice, Enlargement and the Future of Europe* (Manchester University Press, Manchester, 2004)
45. D.M. Farrell, E. O'Malley, J. Suiter, Deliberative democracy in action Irish-style: the 2011 We the citizens pilot citizens' assembly. Ir. Polit. Stud. **28**(1), 99–113 (2013)

46. C. Little, Portrait of a laggard? Environmental politics and the Irish general election of February 2016. Environ. Polit. **26**(1), 183–188 (2017)
47. G. Mullally, E. Byrne, C. Sage, Disciplines, perspectives and conversations. In: *Transdisciplinary Perspectives on Transitions to Sustainability* (Routledge, 2017), pp. 21–40
48. J.O.C.C.A., *Report of the Joint Committee on Climate Action Climate Change: A Cross-Party Consensus for Action.* Dublin: Houses of the Oireacthas, 2019
49. M.T. Boykoff, J.M. Boykoff, Balance as bias: global warming and the US prestige press. Glob. Environ. Change **14**(2), 125–136 (2004)
50. S. Deckard, Introduction: reading Ireland's food, energy, and climate. Ir. Univ. Rev. **49**(1), 1–12 (2019)
51. M. Wahlström et al., Protest for a future: composition, mobilization and motives of the participants in Fridays For Future climate protests on 15, March (2019)
52. N. Gunningham, Averting climate catastrophe: environmental activism, extinction rebellion and coalitions of influence. Kings Law J. **30**(2), 194–202 (2019)
53. D.C.C.A.E., Climate action plan - to tackle climate breakdown (in online) Dublin: *Department of Communications, Climate Action and Environment* (2019). https://www.dccae.gov.ie/en-ie/climate-action/publications/Pages/Climate-Action-Plan.aspx
54. R.L. Ray, V.P. Singh, S.K. Singh, B.S. Acharya, Y. He, What is the impact of COVID-19 pandemic on global carbon emissions? Sci. Total. Environ. **816**, 151503 (2022)
55. H. Daly, J. Glynn, B.Ó. Gallachóir, It is not clear if Green Party 's emissions reduction target is feasible. Online Irishtimescom (2020). https://www.irishtimes.com/news/science/it-is-not-clear-if-green-party-s-emissions-reduction-target-is-feasible-1.4238995
56. E.D. Fraser, A.J. Dougill, W.E. Mabee, M. Reed, P. McAlpine, Bottom up and top down: Analysis of participatory processes for sustainability indicator identification as a pathway to community empowerment and sustainable environmental management. J. Environ. Manag. **78**(2), 114–127 (2006)

57. DCENR, *Strategy for Renewable Energy: 2012–2020* (Department of Communications, Energy and Natural Resources, Dublin, 2012)
58. C. Kuzemko, M. Blondeel, C. Dupont, M.C. Brisbois, Russia's war on Ukraine, European energy policy responses & implications for sustainable transformations. Energy Res. Soc. Sci. **93**, 102842 (2022)
59. G. Gerba, Republic of Ireland 2022. Ir. Polit. Stud. **38**(3), 291–338 (2023)

5

Community and Sustainability—*Dingle Peninsula 2030*

Evan Boyle

5.1 A Specific Place

It is early in the morning on St. Stephens Day, still a few hours left until daylight begins to break. In most places people are soundly sleeping, digesting the biggest of annual feasts. But this is not most places. Here, people are emerging from the cosy warmth of their homes and venturing out into the still cold air of late December, orange streetlights illuminating the dark morning as they head for town. Lá an Dreolín or The Wren's Day (pronounced *wran* locally) is an enduring fascination for the town of Dingle. Although actual hunting of the wren no longer takes place, or even re-enactments of such practices of former times, aspects of the tradition built on community and performativity remain strongly embedded. The festivities consist of four wren groups—Green and Gold Wren, The Quay Wren, The Goat Street Wren and The Sráid Eoin Wren—parading in masquerade around traditional routes of the town. While tradition would dictate that musicians would march in straw 'rigs', conical hats or a mask of some sort, the festivities' garbs have been modernised and anything goes. Glitz and glamour, fancy dress, the

C. McGookin and E. Boyle, *Engaging Communities in Climate Action*, Energy, Climate and the Environment, https://doi.org/10.1007/978-3-031-93498-8_5

macabre and anything else imaginable. From morning to late afternoon, the festivities ensue.

The wren is associated symbolically with treason, having betrayed the first Christan martyr St. Stephen. The tradition of hunting the wren on this day is said to have emerged from this symbolism. While such hunting practices have been relegated to the annals of former times, Wren's Day is still alive and well in west Kerry with the focal point of the tradition a pantomime style hobby horse with a wooden head, snapping jaws and a body made of cloth across a frame. The Wren came close to extinction as a ritual tradition in the last century due to emigrations in the '20 s and '30 s. It also was under threat from the Church who were opposed to the spilling of such celebratory, carnivalesque impulses into public houses and local hotels for much needed refreshments. Despite these existential perils the Wren remains. The traditions are kept going but modernised in places.

The town embraces the celebration—one not only of tradition but of place. Good-natured competition occurs between the Wren groups as people gather for a celebration of community following on from the mainly domestic festivities of the previous day. The Wren Day offers some initial space for reflection on the nature of the town of Dingle and the Peninsula in which it sits. A place steeped in tradition but not bound to it. A place on the outskirts of Ireland, but culturally central. A place where the community shares their connections to a sense of "Dingleness", as insiders in a place so often inhabited by outsiders. Outsiders who are welcomed with open arms and essential to the local tourist industry. However, the insiders are those who rise at 6 am to celebrate with neighbours, and compete with neighbouring streets, aware that this is their place as they strive to sustain their collective lives together in a meaningful, fulfilling and prosperous way.

5.2 Overview of the Dingle Peninsula

To date, exemplar regional energy transitions have tended to be isolated rural areas [See: Chapter 3]. In Ireland, similar challenges can be seen throughout rural areas within the Border, Midlands and Western regions.

The Dingle Peninsula a small isolated 583 km^2 peninsula about 50 kilometres long and 15 kilometres wide, with a population of ~ 12,800 [1], located in County Kerry (dotted line), Ireland, is one such example. Due to the absence of a gas grid and its highly dispersed population, the region spends a significant amount on the import of oil and electricity as in the cases of Güssing and Samsø (See: Chapters 3 and 6) (Fig. 5.1).

Fig. 5.1 Map of Ireland and Great Britain highlighting areas of interest; Co. Kerry (broken line) & Dingle Peninsula (shaded)

Despite, or perhaps because of its isolation, the Dingle Peninsula has established itself as an extremely popular tourist destination. While this is a significant source of income for the area, it brings with it some challenges. The Dingle Peninsula housing stock contains a significant amount of holiday homes at 25% of dwellings compared to 11% in Co. Kerry and 3.2% nationally [2]. In addition, when beginning our study, the limited employment opportunities outside of the hospitality sector and an increasing number of tourists made the area less attractive for young families to settle there. This is reflected in population/demographic statistics, with 41.1% of the population over 50 compared to 30.4% across the state (Fig. 5.2). Culturally, the western side of the area is a Gaeltacht region (a district where Irish is the primary language). It is one of seven officially recognised Gaeltacht regions in the country. Additionally, a Special Protection Area (SPA) dominates the Dingle Peninsula. From just south of Brandon Point in the north, all the way to the end of the peninsula at Slea Head, and as far east as Inch in the south, it includes the high coast and sea cliff portions of the peninsula. The location comprises the sea cliffs, the land that borders the cliff, sand dunes on the Magharees Peninsula and close to Murreagh, as well as several upland regions that are farther inland from the coast. A range of sedimentary rocks, primarily Devonian and Devonian/Silurian sandstones, conglomerates and siltstones, characterise the region. The following species make the SPA of special conservation interest under the EU Birds Directive: Chough, Peregrine and Fulmar. There are distinct features which define the area, both culturally and ecologically, but also in relation to community. Before moving to characterise the community(ties) found in Dingle, we first provide a brief outline of the contested nature of the concept within Sociology.

5.3 Community as a Contested Concept

Definitions of community have long been debated from the earliest sociological accounts provided by Tönnies in Gemeinschaft and Gesellschaft [3] and Weber [4], to theorists of the Chicago School such as Park and Burgess [5]. Despite the long-standing debate on community within

	National		Kerry		Dingle Peninsula			
0 - 4 years	331,515	7.0%	9,020	6.1%	760	5.9%		
5 - 9 years	355,561	7.5%	10,153	6.9%	736	5.7%		
10 - 14 years	319,476	6.7%	9,618	6.5%	740	5.8%		
15 - 19 years	302,816	6.4%	9,106	6.2%	824	6.4%	Dingle	National
20 - 24 years	273,636	5.7%	6,715	4.5%	559	4.4%	28.2%	33.2%
25 - 29 years	297,435	6.2%	7,114	4.8%	547	4.3%		
30 - 34 years	361,975	7.6%	9,722	6.6%	761	5.9%		
35 - 39 years	389,421	8.2%	11,025	7.5%	843	6.6%		
40 - 44 years	357,460	7.5%	10,687	7.2%	914	7.1%		
45 - 49 years	326,110	6.8%	10,418	7.1%	865	6.8%	30.7%	36.4%
50 - 54 years	299,935	6.3%	10,152	6.9%	943	7.4%		
55 - 59 years	270,102	5.7%	9,787	6.6%	987	7.7%		
60 - 64 years	238,856	5.0%	9,156	6.2%	981	7.7%		
65 - 69 years	211,236	4.4%	8,584	5.8%	892	7.0%		
70 - 74 years	162,272	3.4%	6,540	4.4%	637	5.0%		
75 - 79 years	115,467	2.4%	4,427	3.0%	344	2.7%		
80 - 84 years	81,037	1.7%	3,024	2.0%	239	1.9%		
85 years plus	67,555	1.4%	2,459	1.7%	240	1.9%	41.1%	30.4%
All ages	4,761,865		147,707		12,812			

Fig. 5.2 Demographic nationally, in County Kerry, and on the Dingle Peninsula

sociology, an outline is provided here which helps to refine the application of the term to this book. To start, a range of different, yet connected, interpretations of community provided by Walker ([6], p.777–778) provides an insight into the vast rhetorical usage of the concept. These interpretations are indicative of community, rather than comprehensive. Firstly, community is understood as a collective actor with agency that can interact with others. Agency is seen as different forms of capacity which interact with and shape social action in an undetermined manner, in contrast to more functionalist interpretations. In situating community within a stratified, vertical hierarchy, it is understood as a scale of action above the individual or domestic, but below the governmental. Place-related understandings imply a set of social relationships embedded in a geographical locality or territory, e.g. a neighbourhood and a village. Communities as networks can denote social relationships that exist within, but also transgress, geographical boundaries, e.g. communities of interest. When the idea of community is understood as a process, the emphasis is on collaborative, consensual and voluntary involvement in which the quality of social relationships draws on stocks of social capital and trust. Community as identity denotes certain qualities of ways of living, including (self-) representation [7].

Cohen places emphasis on culture over structure in the study of communities [7]. The central question that he is seeking to understand is the way in which people demarcate their immediate and intimate social identities, the boundaries of their social lives which illuminate their sense of similarity and difference. Symbolism is seen as a central process in the creation of boundaries. 'Community', then to Cohen, is premised on two related assumptions, that members of a group share something in common that binds them together, and that this common shared article differentiates them in a significant way from others. The work of Jeremy Brent [8] adds further nuance to debates on community within social theory. Within this work, it is outlined that times of community activity are generally not times of calm, but rather times of turbulence be it political, economic or social: 'community' is taken for granted, until it comes under some kind of threat. Community then is conceived of as a desire, something which is continually replenishing itself as people seek voice and connectedness in all their imperfections. The paradox of community then is the paradox of desire. It is the desire for community, not its achievement, which itself is community. Engaging with community as a phenomenon then becomes an interminable task.

As McNally [9] notes the success potential of Ireland's transition to a low-carbon society depends on the implementation of citizen participation, both in terms of policy support as well as activating citizen engagement in processes of change. The transition pathway cannot be led by governments and utilities alone [10], but must instead be a symbiotic relationship with civil society in which community initiatives and civil society organisations operate to create social cohesion and engagement, with the potential to accelerate the transition by creating a "*favourable context for the implementation of further energy efficiency projects*" ([11]p.5). As Cogan [11] notes, the small group of successful community-led initiatives to date in Ireland have been largely circumstantial, with a need for political commitment to ensure the emergence of such projects becomes normalised. At a grassroots level, community initiatives create an environment in which direct engagement can occur, leading to shared learning [12]. As we have seen with reference to the mainstream media's narrow framing on climate change in Ireland, shared learning processes will need to come from bottom-up engagement, be that community-based or through the work of civil society organisations and NGOs. Citizen engagement can be differentiated from community engagement with the former applying to individuals at the national level and the latter to collective actors with the agency to interact with others.

The grassroots community initiatives to date in Ireland offer an insight into the future potential of such projects to elicit an acceleration of the transition to a low-carbon society. As Brid Walsh [13] suggests, Irish community-led initiatives have gained less attention than their European counterparts. Seyfang et al. [14] reference the emergence of a wide array of community-led initiatives throughout Europe. In Ireland, however, despite being the second-largest per capita CO_2 polluter in Europe ([15] p.226), we have witnessed but a few isolated examples over the last two decades. Cogan [11], in analysing community energy projects in Tipperary and Mayo, suggests how such projects engender high levels of citizen buy-in and can create social contexts conducive to involvement. These projects were stimulated by local needs such as job creation and economic reasoning rather than global environmental concerns, aligning with the general media discourse. There is a need to rethink the 'needs' and 'wants' towards a societal model rather than an individualistic one. People are recruited to such initiatives by others in the community, and continued involvement occurs due to community gains rather than individual gains [16]. The need for the mass mobilisation of community-led initiatives represents the potential to usurp "*dominant individualist and consumerist lifestyles*" which "*run counter to community collectivism*" ([17], p.7560).

In the work of Putnam [18], participation in community work encourages active citizen engagement, which contributes to citizen empowerment, leading to the development of social capital and social cohesion. These can be seen to facilitate trust, and a community will accomplish more if it trusts and is trustworthy. Social capital has been framed as actual or virtual resources which individuals or collective actors acquire through taking part in "*more or less institutionalized relationships of mutual acquaintance and recognition*" ([19], p.119) which allow communities to function smoothly. Social cohesion is seen as "*the willingness of members of a society to cooperate with each other in order to survive and prosper*" ([20], p.5). It can, however, be a somewhat ambiguous concept, which can evoke nostalgic memories for a former time which is deemed preferable or superior, despite a lack of tangible validation for the claim. As such, strong social cohesion can bind communities together but can also accentuate the boundaries as experienced by those excluded. Locally based initiatives can often reinforce existing power inequalities or restrict the scope and impact of participation ([21]). Historically, the "local" has been seen as a key site for building democracy and citizen participation [22]. There is, however, a danger of falling into the "local trap" [23] whereby it is presumed that action and decision-making at the local level are more likely to reap social and environmental benefits. From our starting point concerning the Wren's Day up to the collaborative approach to rural decarbonisation, the question of community is always hovering around. Providing some contextual perspectives on its often-fraught theoretical position within social theory and philosophy can help us move towards situating *Dingle Peninsula 2030* in relation to the concept.

5.4 The Dingle Peninsula as Community

Through the undertaking of my PhD research, I held some initial exploratory interviews with locals. Here, I have included some quotations from locals on the peninsula alongside secondary analysis of interviews undertaken by Hayes-McCoy & Judd in their 2017 book on the culture, history and people of the Dingle Peninsula [24]. These are used to provide insight into the community. Within this, a strong sense of the community spirit, although not explicitly defined, within both the town of Dingle and the villages of the wider peninsula becomes apparent (Table 5.1).

While the cursory notion of community spirit can be outlined, there are further nuances needed surrounding the emergence of different communities of interest on the peninsula. Here, we can invoke some earlier work on communities in the Irish context from Adrian Peace [25],

Table 5.1 Quotes on community spirit

Community Spirit on the Dingle Peninsula
"*The people are shaped by the landscape, the landscape is shaped by the people and there's an instinct there to help and support each other*" ([24] p.8)
"*People in the West depend on each other, and there's a lot of mutual support*" ([24], p.25)
"*You'd have a big sense of community in the villages*" ([24], p.43)
"*I think there are strong community bases there. Not just in Dingle town; Castlegregory and Cloughan and working back towards Dunnquin and Ballyferriter and Ventry, there is network hubs that are already quite active and can get people involved in things*" (Exploratory Interview)

which outlines 3 distinct "communities" existing within one small village demarcated by physical boundaries, social differences, the mechanisms for communication and modes of unification across these established boundaries. Some distinct "communities" which we can outline with relation to their boundaries are the community of Dingle town as distinct from the communities "Back West", following the roads West and Northwest from the town. Within the town, social life is predicated on the interactions on the main street, in shops and pubs, and at the host of regular events held in the town throughout the year. Back West social life is often centred around the parish, the GAA and smaller village centres. The area of the region designated to Irish being spoken as a first language (Gaeltacht Corca Dhuibhne) is interspersed with both speakers and non-speakers. The population is 8,729 with 6,185 Irish speakers [1]; two communities contained within one. Although the sites of social interaction in most cases are the same, there are different modes of communication through which people can converse based upon which community they inhabit. While the Gaeltacht region is geographically defined, it is not the case that this boundary upholds the speaking of the language. Mechanisms of communication are used to decipher the initiated, to the slightly initiated, to the uninitiated. Initial greetings in Irish quickly establish who is fluent and who is not. Often those not comfortable in their own abilities to converse in Irish, despite being quite proficient, are quick to downplay their fluency.

This central importance of language to these two distinct communities within the region played out in a conflict in the 2000s after a 2005 ruling outlawed the use of English place names in the Gaeltacht area. The 700-year-old placename of Dingle was replaced with An Daingean, with signage officially updated throughout the town. A campaign to reinstate the name Dingle emerged with unofficial updates made to place the name Dingle back on road signs. A compromise was reached in 2011 whereby a bilingual description of the town was settled upon. An Daingean has been replaced with the more traditional Daingean Uí Chúis, and the town name decided as Dingle/Daingean Uí Chúis. The two communities in one reached a consensus in part due to the importance of Dingle as an international brand for tourism. The recency of these debates highlights the continued contestation surrounding meanings of place.

In thinking about language we see just a few insights into the multiple ways in which communities of interest can manifest in a distinct location, continuing on from the work of Peace [25]. Others can be outlined with relation to employment type (Fig. 5.4) or locals and blow-ins, or with regard to Gaelic Football between An Ghaeltacht, Annascaul, Castlegregory, Dingle, and Lios Póil. The peninsula has long since been an area of attraction to artists with both locals and those who move there being inspired by the region. On from this, more demographical distinctions can be made. Women of the Dingle Peninsula and men of the Dingle Peninsula, young people and elderly people, or differing socio-economic distinctions. Within all these significant notions of difference we must ask if a unifying community could ever exist? Peace [25], however, goes against some notions at his time of writing around the decline of community, instead in his case study context finding no such decline but instead three distinct communities which come together in spite of their differentiation through notions of outsiders and the role of cooperative associations. Some similar modes of unification can be outlined regarding the Dingle Peninsula.

The creation of the brand of "Dingle" is something that plays a unifying role, similar to the discussion of Orkney Ltd. found within Chapter 3. It comes to be defined as "our" place by locals, a place in which tourists can come to visit for a small window in the year, a time in which employment in the tourism sector booms, before reverting to normal life. The roads play a unifying role. Both the Slea Head Drive and the Connor Pass are central passages from the nucleus of Dingle Town west to Ballyferriter and north to Brandon and are shared spaces of natural beauty. These passageways are a regular point of conversation, either in passing or at length and transcend many of the boundaries created through distinct communities of interest. On from this, the physical conditions which people on the peninsula must endure (or embrace as is more often the case) act as a mode of unification. This hardness of life is something that is represented through the landscape, through the history, and through the inclement weather made more severe by the mountainous nature of the region.

While offering some outlines of the modes of unification that may envelop the distinct and bounded communities of interest, the diverse forms that community takes are important considerations when discussing notions of community engagement and community-led projects more broadly. This is a complex and ethically subtle situation in which I must reflexively interpret my own positionality through the process as outlined in Chapter 2 on transdisciplinary approaches to research. I have sought to gain a good grounding in different dynamics on the peninsula to inform my approach to the research, and my own place within it. In initially undertaking the work I boldly laid claim to my plans to learn the language, before feebly settling for cúpla focal. In some contexts, for example with those individuals strongly committed to the Irish language, I held my nerve for as long as possible in introductory conversations before reverting to English to explain my limitations and outlining my genuine appreciation for the language and my desire to learn it at 'some stage'. This often led to discussions about the role of the secondary education system in non-Gaeltacht regions. Within the collaborative group, a decision was made to use the name *Corca Dhuibhne 2030/Dingle Peninsula 2030* in light of the conflicts outlined above.

The Dingle Peninsula, as like any region, is full of complexity with regard to the notion of community. Diverse and bounded communities of interest exist while some overarching modes of unification are also present. Engaging in research collaborations on the peninsula, both Connor and myself needed to remain aware of these be they Dingle Town or Back West, Irish speaker or non-Irish speaker.

5.5 Emigration, (un)Sustainability, and Employment

While a sense of community is strong within the peninsula, there is also widespread emigration, with a history dating back generations and most acutely felt in the middle of the nineteenth century. Today, young people leave the peninsula for work or to attend college in cities in Ireland or further afield in America, Australia or the United Kingdom, often never to return. This emigration pattern has been acknowledged as a key concern in the analysis of the secondary data found in the work of Hayes-McCoy & Judd [24]:

> One of the biggest changes here is the number of people who are left in the villages. When I was growing up if you walked through a village there were always people outside doing things, kids playing on the roads…emigration I suppose has changed that ([24] P.107).

Looking quantitatively at the area, as of 2016 prior to our investigation, 41% of residents were aged above 50, well above the national average of 30%. One of the most striking demographical insights is that the population on the peninsula aged 20–29 was 9% compared to the national average of 12%. While some of this difference can be attributed to the lack of a third level institute on the peninsula (the United States Sacred Heart University does, however, have a campus in Dingle), it also suggests that young people are not returning to the area on completion of their education.

During An Gorta Mór, the Great Famine, many of the townlands and villages of the peninsula were devastated both by emigration and starvation. Many of these villages have not since recovered, with the town of Dingle gaining prominence through the twentieth century as the nucleus of the peninsula. Many villages lost close to half of their population during this time with Ballyferriter (-50%), Dunquin and the Blasket Islands (-48%) and Minard (-48%) most effected, followed on by Kilquane (-43%), Kilmalkedar (-43%), Garfinny (-44%), Kinard (-43%) and Ventry (-42%) (Census 1841–1851). Emigration has a cultural precedent on the peninsula, and although it occurs today due to less severe circumstances than those of the 1840s, it is still concerned with people leaving the peninsula in search of a perceived improvement in the standard of living (Fig. 5.3).

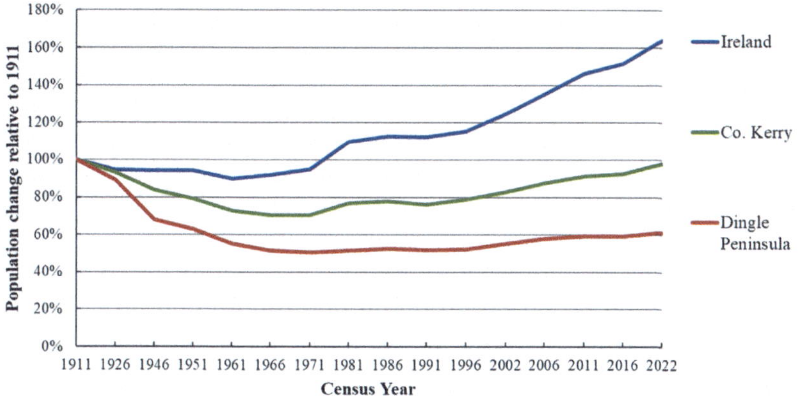

Fig. 5.3 Population change in Ireland, Co. Kerry and the Dingle Peninsula from 1911 to 2022

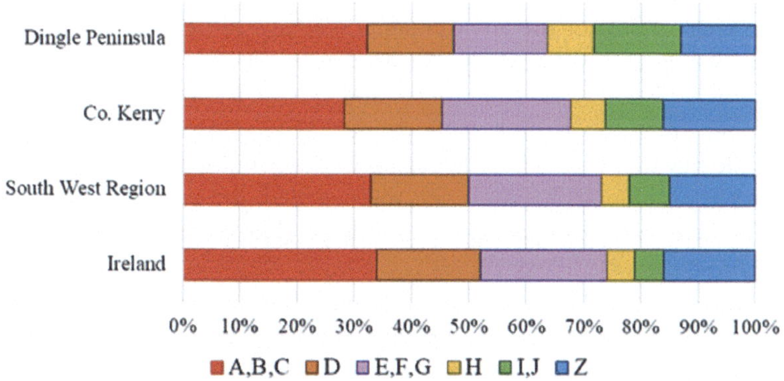

Fig. 5.4 Social Class Composition for employment for 2016. A Employers and managers; B Higher professional; C Lower professional; D Non-manual; E Manual skilled; F Semi-skilled; G Unskilled; H Own account workers; I Farmers; J Agricultural workers; Z All others gainfully occupied and unknown ([27] P. 20)

While emigration is an issue, it is also seen as an opportunity to create the structures to enable people to return to the peninsula, as illustrated through analysis of the initial exploratory interviews:

> I also think there is such a strong affiliation with the area here I think everyone who is from Dingle really wants to come back here, or wants to see it being successful so that is one strength. I think if we put the eco system in place for people to come back here, they will come back because if you give them the opportunity there are many many people who really would like to... I don't think you can say that about everywhere. (Exploratory Interview)

Another concern outlined through the exploratory interviews, and linking back to the earlier discussion on tourism, is the need for sustainable tourist strategies to preserve both the environmental and social strengths of the area.

> I think people realise on the peninsula, especially with the growth of tourism in the past decade, the importance of the environmental issues. That we have to be seen to be proactive on environmental issues. We have to be seen to protect the environment. We have to be seen to be using energy wisely and water wisely. We get lots of talk of congestion on the peninsula. There are so many buses and cars and is there a better way of doing things. So I think some people are aware that tourism is very valuable but it also could be damaging to the peninsula. It gives you the wrong impression, or it simple gets too crowded, the carbon footprint of getting here and getting around here is high. (Exploratory Interview)

The relationship between the area and the environment is explored through the *Dingle Peninsula 2030* project, with specific relevance to decarbonisation and the creation of sustainable solutions to developing the peninsula moving forward.

On the western coast of the peninsula, at Ferriter's Cove, some of the earliest archaeological evidence for agriculture in Ireland have been found, dating back to c. 4,000 B.C. [26]. Since then, the area has had a long and storied relationship with labour and employment. Throughout the sixteenth century, the port in Dingle town operated as an important hub for merchant ships trading goods with France, Spain and Britain. Throughout the nineteenth century, Dingle was one of the poorest and most isolated regions in the country, heavily reliant on subsistence farming, culminating in the devastation caused by the Great Famine. *Dingle Peninsula 2030* is centred on the creation of a sustainable future for the peninsula, with employment seen as central to this. Historically, however, the area has been defined by unsustainability in different forms. The colonial creation of plantations throughout much of the east and midlands of the country pushed displaced peasants west towards places like Dingle from the sixteenth century onwards. This was not the only means through which the Peninsula was impacted by the occupation of Ireland, with the area suffering damage and destruction through both the Nine Year's War (1594–1603) and the Wars of the Three Kingdoms (1639–1653) during the reign of Charles I.

In the contemporary case, as outlined by Ó Caoimh and McGookin [27] the Peninsula has a lower proportion of high social class categories of employment than the national average. There are limited manufacturing-based employment opportunities leading to a small number of people in occupations classified as manual and semi-skilled (See: Fig. 5.4). Relatively speaking there is a higher percentage of 'own-account workers', related to the tourism industry and there is a higher significance of both agriculture and fishing. Both sectors, however, are situated within unsustainable current practices and historical timelines which have led to the contemporary situation. There is a heavy focus on dairy and cattle on the peninsula, due not only to topography but to national policy and its ever-changing nature. The support of grants and schemes is essential to make farms economically viable. Fishing in the region, as with much of the world, has been impacted by the intensification and globalisation of activity, leading to the degradation of Irish fisheries and global marine destruction alongside the societal impact of the evaporation of these established communities.

The high degree of "own-account workers" predominantly associated with the tourism sector also draws out aspects of the unsustainable nature of the region. Predicated on its natural beauty, tourism in Dingle suffers from negative practices with relation to bus tours, air travel, holiday homes, and Airbnb. Its seasonal nature means that employment is often precarious, yet labour intensive in the busy season with people working long hours without days off for weeks and sometimes months on end. In seeking to investigate *Dingle Peninsula 2030* and its vision for a sustainable future, an awareness of the unsustainability of the region is necessary, both to situate where a desire for this may stem from and to help frame any suspicion which may exist towards claims of salvation.

5.6 *Dingle Peninsula 2030*

Our collective research on the Dingle Peninsula followed an action-research approach, with some brief reflections needed on how this informs our work as part of *Dingle Peninsula 2030*. In discussing the long-term vision underpinning continued commitment to action research, Brydon-Miller et al. ([28], draw on trust-building within community contexts where academic outsiders have often, and often rightfully, been treated with caution. As they note "*a key value shared by action researchers, then is the abiding respect for people's knowledge and for their ability to understand and address the issues confronting them and their community*" ([28], p.14). This idea of respect for local people and their knowledge has poignancy in the context of the Dingle Peninsula, where anthropologist Nancy Scheper-Hughes completed the field research for her 1979 publication *Saints, Scholars, and Schizophrenics*. While the book gained academic success within the fields of psychological and medical anthropology, it was viewed by the Press in Ireland, and the people of the pseudonymised village of 'Ballybran', negatively, with its publication seen as an unethical violation of privacy. Some twenty years after the initial scandal, Scheper-Hughes returned to the Dingle Peninsula and the town of 'Ballybran', only to be exiled from the village a few days after her arrival [29]. Although the context of Scheper-Hughes' work, an anthropological investigation into sensitive psychological topics, is far removed from *Dingle Peninsula 2030*, the lessons learnt give further credence to the need for trust-building and mutual understanding when pursuing an engaged or action-research approach.

There are, however, further nuances to the approach taken. Within Shaper-Hughes' work, the perceived unethical violation of privacy had an impact on the work. Our approach sought to balance the tension of being an action researcher both within the collaboration and as an outsider to the community. Seeking too much distance from the subject matter, or alternatively becoming too embedded as a "project partner" rather than as a researcher can lead to problems. We sought to work in close collaboration with the *Dingle Peninsula 2030* project in order to have a positive impact on the process under investigation while at other times took a detached perspective in order to undertake the construction of analysis on the situation that was both being investigated and which we were collaborating on. Understanding how to manage this movement between inside and outside is held throughout the process of the research project and has been outlined previously (See: Chapter 2).

The group was predicated on a commitment to working collaboratively. Despite the organisations represented having never worked together, the collaborative committee developed a working style which enables all voices to be heard. On some decisions one or other of the groups may have taken the lead, but all members contributed their views. Over the time frame relevant to our direct involvement (2018–2021), the group was engaged in regular meetings (29 meetings) and continuous email interactions. While collaboration has been embedded in the group, reflection was an emergent practice developed due to the need to document and learn from our experiences. This need emerged from the group through conversations, as the overall project began to develop and grow.

As collaborative projects are creative, sometimes experimental and fraught with inevitable and unexpected structural, organisational, and process challenges, almost all aspects of the process must be open to continuous re-examination and re-evaluation [30]. Feedback can then be used to strengthen the collaborative relationships and their effectiveness [31]. We reflected on the collaborative experience we have had, engaging together in transdisciplinary research practice, "*transgress(ing) boundaries between scientific disciplines and between science and other societal fields … include(ing) deliberation about facts, practices and values*" ([32], p. 5).

5.7 Concepts of Community in Context

A central facilitator and driving force in *Dingle Peninsula 2030* was Dingle Hub (Mol Téic), which operates as a not-for-profit entity and a community enterprise initiative (2018–ongoing), using technology to bring employment opportunities and business start-ups to the peninsula, utilising the 1 GB fibre connection installed. Key partners working with the hub through this period included Eir (national broadband provider), Dingle Business Chamber, Údarás na Gaeltachta (regional authority responsible for the economic, social and cultural development of the Gaeltacht), Kerry County Council and Net Feasa (Internet of Things business on the peninsula). The hub, on inception, aimed to facilitate the creation of 100 new jobs within five years. This was noted by a board member:

> the first thing was to see if we can create employment, the concern when you looked at the demographics and everything and Dingle was an aging community and Kerry was aging but Dingle in particular and young people, any of those locally were in seasonal employment generally poorly paid, and that was in tourism (Exploratory Interview)

As well as the employment aims, Dingle Hub was established to be a centre for education, partaking in the 2018 University College Cork (UCC) University-Wide Module on Sustainability. The module was taken by over 20 people on the peninsula who met at the hub each week to digitally attend classes facilitated by UCC. This highlights the potential of digital connectivity for rural areas in relation to education. An outline of the plans around education for Dingle Hub were detailed on their website as such: due to "*the importance of education and the need to provide the educational skills and knowledge to enable people to prosper and thrive in what is now being called the 'Fourth Industrial Revolution' (i.e. the interconnected world of people and things and the growing importance of data), it is planned to develop an educational programme about the issues involved in such a transformation and that will be based in Dingle*".

Dingle Hub acted to develop *Corca Dhuibhne 2030/ Dingle Peninsula 2030*) as a collaborative project aiming to transition the region to a low-carbon, sustainable community by 2030, working across scales, including grassroots community organisations, regional authorities and national agencies in developing and delivering a range of sustainability projects across the sectors of marine, agriculture, energy, tourism and transport. Grassroots innovations, within transitions research, are outlined as differing "*from market-based innovations because they foreground moral values and collective aspirations, aim to change social practices and 'ways of living', are highly contextual, and are often developed in response to local problems*" ([33], p. 193). *Dingle Peninsula 2030*, although not strictly an example of a grassroots innovation due to the multi-level collaborative governance and partnership at its core, aligned with some of the characteristics of such. Dingle Hub operated in a facilitating role between the different actors, organisations and community representatives involved in the range of different initiatives under the umbrella of *Dingle Peninsula 2030*. Acting both as facilitator, and in many cases initiator, Dingle Hub sought to support a transition aligned with the initial vision set out for *Dingle Peninsula 2030*. The key objectives of the project were to work with the local community, schools, business and farming sectors to explore, support and enable the broader societal changes required for the low-carbon transition, while also working to identify employment opportunities and to help create the conditions for a sustainable economic future for the area. The collaborative committee was central to the operations of the project and the approach taken has acted as an innovative means through which the merging of top-down and bottom-up approaches can be investigated.

5.8 Concluding Remarks

Continuing from the previous discussion on the contested nature of community, the Dingle Peninsula is defined firstly as a geographical place-based community. The community is made up of numerous symbolic boundaries, or contested publics, defined by parishes, sports clubs, culture, language, local activities, amenities, labour types, etc.

The main areas in the region are Dunquin, Ballydavid, Ventry, Dingle, Cloghane, Castlegregory and the Maharees, Camp, Annascual & Inch, and Lispole and Minard and while community within these diverse areas are established through the symbolic construction of boundaries there still lies a unifying sense of community that operates across the whole peninsula. This is not, however, to suggest a homogenised culture, but rather an overarching frame of community to which most inhabitants relate. As noted, the area is a popular tourist destination and has a large amount of holiday home owners, adding further nuance to notions of community such as an extremely busy seasonal trade across hospitality sectors and tourism when compared with the much quieter winter months when many permanent residents move to unemployment or part-time work. This has influence on the social dynamics of daily life across the seasons of the year.

The *Dingle Peninsula 2030* project is borne out of a notion of community as desire, as per the previously outlined work of Brent [8]. The desire relates to rectifying some of the issues of unsustainability as outlined earlier in this chapter, be that in relation to employment, energy use, transport, food systems, and so on. Desire is firstly predicated on the goal of bringing sustainable employment to the peninsula and on from this the vision of a sustainable community across a range of sectors. Continuing from the work of Brent, community in this context operates as a desire for something which may never, in fact, be achieved but it is in the pursuit of this desire that community is formulated, stabilised and continued. *Dingle Peninsula 2030* was predicated on this desire for community, the desire for a sustainable community that is more self-sufficient and which can provide careers for people in the local area in order to prevent further emigration from the region. The project was dependent on this desire for community and seeks to enable the emergence of this desire among other people in the area. As Brent notes, this desire is what constitutes community rather than any defined endpoint, despite connotations to time contained within the outline of *Dingle Peninsula 2030*. The project, although unified on some level by a shared vision towards decarbonisation, created the space through which people could envisage their own desired future for the community and in doing so draws people together to work on the delivery of different projects, initiatives and goals.

References

1. Central Statistics Office, Home/Census/Census 2022/Census 2022 Small Area Population Statistics. (2023). https://www.cso.ie/en/census/census 2022/census2022smallareapopulationstatistics/. Accessed 18 April 2024
2. Central Statistics Office, Census of Ireland. (2016)
3. F. Tonnies, C.P. Loomis, *Community and Society* (Routledge, London, 2017)
4. M. Weber, *Economy and Society: An Outline of Interpretive Sociology* (University of California Press, Berkeley, 1978)
5. P.D. Goist, City and" community": the urban theory of Robert Park. Am. Q. **23**, 46–59 (1971)
6. G. Walker, The role for 'community'in carbon governance. Wiley Interdiscip. Rev. Clim. Change **2**, 777–782 (2011)
7. A.P. Cohen, *Symbolic Construction of Community* (Routledge, London, 2013)
8. J. Brent, The desire for community: Illusion, confusion and paradox. Commun. Dev. J. **39**, 213–223 (2004)
9. B. McNally, Media and carbon literacy: shaping opportunities for cognitive engagement with low carbon transition in Irish media, 2000–2013. Razón y Palabra **19**, 119–150 (2015)
10. G. Mullally, E. Byrne, A tale of three transitions: a year in the life of electricity system transformation narratives in the Irish media. Energy Sustain. Soc. **6**, 1–14 (2016)
11. S. Cogan, Community-led energy initiatives in ireland: accelerating the energy transition? Reflections on the impact and outcomes of two case studies. Ir. J. Soc. Econ. Environ. Sustain. **1**, 4–21 (2017)
12. S. Moloney, R.E. Horne, J. Fien, K. Morgan, Transitioning to low carbon communities—from behaviour change to systemic change: lessons from Australia. Energy Policy **38**, 7614–7623 (2010)
13. B. Walsh, Community: a powerful label? Connecting wind energy to rural Ireland. Commun. Dev. J. **53**, 228–245 (2018)

14. G. Seyfang, J.J. Park, A. Smith, A thousand flowers blooming? An examination of community energy in the UK. Energy Policy **61**, 977–989 (2013)
15. E. Fox, H. Rau, Disengaging citizens? Climate change communication and public receptivity. Ir. Polit. Stud. **32**, 224–246 (2017)
16. S.M. Hoffman, A. High-Pippert, From private lives to collective action: recruitment and participation incentives for a community energy program. Energy Policy **38**, 7567–7574 (2010)
17. L. Middlemiss, B.D. Parrish, Building capacity for low-carbon communities: the role of grassroots initiatives. Energy Policy **38**, 7559–7566 (2010)
18. R.D. Putnam, Bowling alone: America's declining social capital, in *Culture and Politics*. ed. by L. Crothers, C. Lockhart (Palgrave Macmillan, New York, 2000), pp.223–234
19. P. Bourdieu, L.J. Wacquant, *An Invitation to Reflexive Sociology* (Polity, Cambridge, 1992)
20. D. Stanley, What do we know about social cohesion: the research perspective of the federal government's social cohesion research network. Can. J. Sociol./Cahiers Canadiens de Sociologie **28**, 5–17 (2003)
21. P. Parvin, Against localism: does decentralising power to communities fail minorities? Polit. Q. **80**, 351–360 (2009)
22. J. Gaventa, E. Jones, A.I. Gaziulusoy, C. Ryan, S. McGrail, P. Chandler, P. Twomey, Concepts of citizenship: a review. Development Research Centre on Citizenship, Participation and Accountability (IDS). J. Clean. Prod. **123**, 55–64 (2002)
23. M. Purcell, J.C. Brown, Against the local trap: scale and the study of environment and development. Prog. Dev. Stud. **5**, 279–297 (2005)
24. F. Hayes-McCoy, W. Judd, *Dingle and its Hinterland* (Gill & Macmillan Ltd, Dublin, 2017)
25. A.P. Cohen, *Symbolising Boundaries: Identity and Diversity in British Cultures* (Manchester University Press, Manchester, 1986)
26. P.C. Woodman, E. Anderson, N. Finlay, *Excavations at Ferriter's Cove 1983–1995. Last Foragers/First Farmers in the Dingle Peninsula* (Wordwell, Dublin, 1999)
27. B. Caoimh, C. McGookin, Dingle Peninsula Community Meetings February 2020, 2021, (n.d.)
28. M. Brydon-Miller, D. Greenwood, P. Maguire, Why action research? Act. Res. **1**(1), 9–28 (2003)
29. N. Scheper-Hughes, Ire in Ireland. Ethnography **1**, 117–140 (2000)

30. N.C. Roberts, R.T. Bradley, Stakeholder collaboration and innovation: a study of public policy initiation at the state level. J. Appl. Behav. Sci. **27**, 209–227 (1991)
31. L.R. Bronstein, A model for interdisciplinary collaboration. Soc. Work **48**, 297–306 (2003)
32. U. Wiesmann, S. Biber-Klemm, W. Grossenbacher-Mansuy, G.H. Hadorn, H. Hoffmann-Riem, D. Joye, C. Pohl, E. Zemp, Enhancing transdisciplinary research: a synthesis in fifteen propositions, in *Handbook of Transdisciplinary Research*. ed. by G.H. Hadorn, H. Hoffmann-Riem, S. Biber-Klemm, W. Grossenbacher-Mansuy, D. Joye, C. Pohl, U. Wiesmann, E. Zemp (Springer, Dordrecht, 2008), pp.433–441
33. F.W. Geels, Socio-technical transitions to sustainability: a review of criticisms and elaborations of the multi-level perspective. Curr. Opin. Environ. Sustain. **39**, 187–201 (2019)

6

Developing a Community-Led Strategy: Co-Production Approach to Energy Planning

Connor McGookin

6.1 What Is an Energy Plan?

A vital first step in the regional energy transition process is to develop an understanding of the current energy balance, detailing the energy demand/supply in each sector and associated CO_2 emissions [1]. This is highlighted by the requirement when joining the Covenant of Mayors to commit to the development of a Baseline Emission Inventory (BEI) as a pre-requisite for the creation of a Sustainable Energy & Climate Action Plan (SECAP) [2]. The Covenant of Mayors for Climate & Energy was launched in Europe in 2008. Following its success in Europe, it was later followed by the Global Covenant of Mayors launched in 2015 [3]. Although it was initially intended to simply cover the EU's largest pioneering cities, it has now gathered widespread support, and by the end of 2024, had over 12,000 signatories of which 8339 have prepared an energy plan [4].

An energy plan/strategy can be developed for any geographic area: countries, regions, cities or smaller towns and villages. It generally has two components: 1) a baseline establishing what fossil fuels are currently

C. McGookin and E. Boyle, *Engaging Communities in Climate Action*, Energy, Climate and the Environment, https://doi.org/10.1007/978-3-031-93498-8_6

needed for, and 2) a strategy to reduce fossil fuel use. Within the baseline, there is usually:

- **Energy demand**—An estimate of what is currently driving energy demand, it is often referred to as energy service demand, i.e. what we get from the energy. People generally don't burn fossil fuels for no reason, they need warm or cooled homes, lighting, cooking, travel, as well as access to goods and services that must be produced.
- **Energy supply**—This outlines the current fuels being used. How are the demands for energy services being met? It will provide a breakdown for each sector of the fossil fuels used, which can then be used to determine how much is spent on fuel and the associated CO_2 emissions.
- **Energy balance**—An energy balance brings demand and supply together to show how energy flows from one to the other.

Common components in the strategy will be:

- **Energy efficiency improvements**—The abundance of cheap fossil fuels has resulted in an energy system that is very wasteful. There are huge opportunities to reduce fuel demand by improving the energy efficiency of buildings. More recently, there has also been a growing interest in ways to reduce demand such as switching transport modes or developing more compact housing.
- **Renewable energy assessment**—This will outline what local resources are available and how much energy they can provide. It may go as far as identifying potential sites for renewable energy projects such as wind, solar or bioenergy.

An energy plan only captures part of the picture. More comprehensive climate action plans or climate mitigation strategies will also include non-energy emissions. A complete baseline emissions inventory will include other greenhouse gas sources such as methane (CH_4) and nitrogen oxide (NO_2) from land use, waste, agriculture and industry.

6.2 Creative Approaches for Involving the Community in Energy Systems Modelling

To develop my approach, I performed a systematic review of the academic literature on energy system modelling processes that included integrating participatory methods to involve different groups from outside the research team [5]. The main focus was on the methods used and how the different projects had linked the primarily qualitative inputs from participants with the quantitative energy system modelling exercise. There were lots of examples of conventional methods such as interviews, surveys, and workshops. However, I was drawn to approaches like serious games or simulation tools that demonstrate the energy system challenges through interactive exercises.

A serious game is an interactive approach to facilitate discussions on complex problems and support learning across a diverse group of actors [6, 7]. A well-known example is the World Climate Simulation game, which is a role-playing game that involves people negotiating as different country representatives in a mock UN climate summit [8]. It comes with an online tool, C-ROADS and En-ROADS, to show participants the impact of their negotiations on the climate. There has been a variety of serious games developed in energy systems modelling studies. In general, these involve the participants selecting a portfolio of energy technologies or adjusting levers/sliders to vary energy scenarios. This has been done through the use of role-playing [9], maps [10–12], and computer applications or online dashboards [12–19].

The use of maps and role-playing offers a means to engage with two prominent drivers of public opposition to new energy developments: concerns about landscape changes and perceptions of being excluded from planning decisions. Pia Nabielek, Hartmut Dumke, and Kurt Weninger developed a very innovative approach that involved pining pieces of card onto a map of the area, with the size of the card scaled to represent the technologies energy potential (kWh/m^2/year) [11]. Participants were given a booklet containing background information on the existing energy facilities, energy demand of the region and a summary of renewable energy technologies. This allowed the participants to step into the role of energy planners and engage with the land use trade-offs necessary to meet the regional energy demand with local renewable sources. In another interesting example from Switzerland, Franziska Steinberger, Tobias Minder, and Evelina Trutnevyte used the combination of an interactive dashboard and map-based board game [12]. Participants first selected a preferred portfolio of renewable energy using a computer application and then went on to place them on a map. Similar to these approaches was the use of energy proposal cards for a fictional town, providing information on the potential sites for power plants and the associated contribution each would make to jobs or climate targets [9]. Participants were then asked to assume the role of local decision-makers (in the form of 'town councillors' or 'council members') and rank the proposals.

Interactive computer tools or dashboards are also a good way for people to take on the role of an energy system modeller. This approach generally involves providing people (often members of the public) with information on different renewable energy technologies and then asking them to choose a portfolio of technologies to meet a certain energy demand. The dashboard or interface differs according to the approach being used. Christina Demski, Alexa Spence, and Nick Pidgeon built a web-based tool that allowed users to explore their preferences for different supply and demand options by displaying the changes on an animation at the level of home, city, and country [19]. After making adjustments to achieve a CO_2 reduction target, users had the option to submit their scenario to a research database. There are several other studies that have involved using a dashboard to select an electricity portfolio, which came in the form of a simplified excel representation downloaded from an online portal [15, 16], or was provided during a workshop session [17, 18, 20, 21]. The primary focus of these dashboards is generally energy and CO_2 emissions; but one tool also displayed the impact of chosen technologies on land, water, and health [16]. Three of the studies provided factsheets to the participants to help them assess the importance of environmental, health, or economic impacts [14, 16, 18]. The novel work of Xexakis et al. compared the difference between 'informed' citizens given factsheets and a sample that were given no supplementary information [17].

These games can help raise awareness, build capacity for problem-solving, and provide a useful space to explore complex problems [22]. It gives people the tools to see in real-time their energy system configurations and how it will impact key factors like CO_2 emissions or land use trade-offs. However, this generally comes at the cost of simplifying the energy system model or selecting a small group of results to share.

6.3 Community Engagement Process to Develop *Dingle Peninsula 2030* Energy Scenarios

It was expected that the community energy group (Dingle Peninsula SEC) formed by our local partners to coordinate energy efficiency and renewable energy projects would see a long-term planning exercise as the logical next step having determined a baseline for current energy-related CO_2 emissions. I prepared a research plan based on the studies summarised in Sect. 6.2, which was to follow a three-stage engagement process: a scenario visioning exercise, discussion on pathway or technology preferences, and finally an evaluation/feedback session. The iterative nature of the process was seen as important to give those participating an opportunity to see how their inputs were being used in the model. In existing energy system modelling studies, it is rare that participants are given the chance to evaluate how their input informed the results of the modelling exercise [5, 23].

We had several meetings within the research team to discuss and refine the research plan. This was very new territory for the organisation (MaREI), so there was pressure to have a clear picture of what our involvement in the Dingle Peninsula would look like and demonstrate why the project should get the extra funding that it was drawing. However, it was difficult to get buy-in from our local partners. There were several reasons for the delay and eventual scrapping of the original research plan. While our local partners saw clear benefits to collaborating as part of *Dingle Peninsula 2030*, they were hesitant about opening up to the wider public or bringing in other community groups. There were concerns around how the decision-making process would evolve. Those working with us, did so on top of their day jobs, and therefore had limited capacity. Dingle Hub thus preferred to select a couple of key projects it could work to deliver within the next 12–24 months rather than open up discussion on a broader strategy for the region over the next decade beyond the high-level vision of *Dingle Peninsula 2030*.

As an educational resource, the original research plan along with the guiding principles from transdisciplinary research that shaped it are presented here before outlining what was actually delivered in terms of community engagements. The focus of this chapter is on the energy system modelling exercise conducted to prepare scenarios for 2030, and how the different engagements informed that approach. How the actual engagements were run is covered in detail in Chapter 8.

Guiding Principles

The engagement of different groups in research through approaches like community-based participatory research or other action-orientated approaches is not a new practice (Chapter 2). I had undertaken a module in community-based participatory research, which emanates from the health sciences, and also attended a four-day training course on facilitation hosted by a community development organisation. The emphasis within these practices was very much on a participant driven process. This stems from the fact they are generally dealing with vulnerable communities. There is also great concern for the danger of 'helicopter research' and the use/abuse of research participants to fulfil a project goal. I wanted to follow this best practice principle, although it would mean going a step further than most other energy planning exercises. The guiding principles can be summarised as the following three pillars:

1. Establish a committee/partnership—Following the example of Jühnde (Chapter 3, Sect. 3.4), and other transdisciplinary projects

[24], the research team setup and coordinated a collaborative governance committee to oversee activities in the area, and facilitate input into the research design and implementation process. This committee was comprised of seven members: a board member and the manager of the Mol Téic (Dingle Hub, enterprise development organisation), the Local Area Manager within NEWKD (community development organisation), Ireland's distribution system operator (ESB Networks) Dingle Community Engagement Manager and the MaREI research team. The committee met regularly on a monthly or sometimes bi-monthly basis to discuss ongoing projects and coordinate joint activities.

2. Flexible and adaptable research design—A core component of the approach taken was that it was flexible and great effort was made to design the research process with our local partners, evolving along with activities on the ground. Following best practice in transdisciplinary research, representatives should have the opportunity to input on research design as well as shaping the results [5].
3. Contribute to the local efforts—The research should foster real-world impact. I had two main contributions.

    a. I was a member of the steering committee setup to oversee the development of an initial estimate of the Dingle Peninsula's energy-related CO_2 emissions, provided the necessary analysis, and co-produced a report on it with our local partners [25].
    b. The community planning process that I was involved with was not only coordinated in partnership with NEWKD but also co-funded by the research centre.

Original Research Plan

This section presents the one-page summary I prepared to share my research plan with others in MaREI's *Dingle Peninsula 2030* team. It would have followed a three-stage process to develop energy scenarios for the region.

<u>Stage 1: Interviews, Autumn 2019</u>

Initial semi-structured interviews to understand local perspectives on the future of the Dingle Peninsula. These should be broad discussions on people's concerns and the challenges facing the community, not solely focused on climate change. The purpose of these conversations is to develop storylines for the future of the region that will be used to create scenarios in the energy system model. This should also capture preferences for benefits from projects like the ESBN or Dingle Peninsula SEC initiatives. Through contacts in the area key organisation representatives will be a starting point. However, in the interest of capturing the wider community, other approaches may be taken such as attending some events in the area or standing outside the church, shops, post office, or attending sports events.

Potential Questions;

1. What's your vision for the future of the Dingle Peninsula?
    - How do you think the region will change over the next few decades? What changes would you like to see?
    - What sort of Dingle Peninsula would you like to see in 2030? What has to change now for that to be realised?
    - The Dingle Peninsula has a very strong sense of community, how do you see this developing in the coming years?
2. What energy options do you have in terms of heating/transport? What improvements could be made?
    - I found it difficult to get out here; have you ever had issues with getting around?
    - How do you heat your home?

Stage 2: Scenario Development Workshop, Winter 2019/Spring 2020

Present findings from stage one and discuss potential scenarios to be explored. These should place energy and climate change within the local context, and identify opportunities for rural development.

Potential Exercises:

- 1st half: develop concept map—begin by feeding back the various concerns and opportunities gathered from meeting the different groups and then look at what the vision for the future is and the path to get there.
- 2nd half: two scenarios presented 1) following the national ambition and 2) energy independence, participants rate technologies and then explore how they would place them on a map of the area.

Stage 3: Feedback Workshops, Autumn/Winter 2020

Presentation of the modelling results at public events. The annual Farmer's Forum would be a good opportunity as it is well established and receives strong attendance. There should also be dedicated presentations to interested groups (e.g. SEC, energy mentors and ESB ambassadors) either together or as a small event. The purpose of the exercise is to gather feedback on the modelling results and what they mean for the community, identify key projects to work on, as well as exploring how well they feel the scenarios reflect their concerns and priorities. It is important to stress that these are draft results and there will be a revision to the model to address the feedback received.

Revised Research Plan

In collaboratively agreeing a community engagement process with our local partners it became clear that the framing of discussions would need to be wider than just desirable climate and energy futures. The research became part of a broader effort led by the local development organisation (concerned with establishing a baseline against which socio-economic and demographic trends could be assessed) to align with the community's needs. This was seen as important: to avoid '*preaching to the converted*' and reach a more diverse group, as well as getting a more comprehensive understanding of the region. There were two rounds of community meetings, with a total of 15 meetings across the 8 parishes that make up the region. During the first round of community meetings, participants were essentially given a blank page to outline their concerns and priorities. This posed a serious methodological challenge for the modelling but was a required trade-off to stay true to the participatory process.

A brief summary of the community engagements is provided in Table 6.1 (for a detailed description please see Chapter 8 Sect. 8.3.1, 8.3.2, 8.3.3 and 8.4.1). It had four main elements. Firstly, the initial scoping exercise with university students from the case study area. This involved several informal meetings before bringing a group of thirteen together for a workshop discussion. During the discussion a series of questions were posed before grouping the answers under common themes. Secondly, there were two rounds of community meetings held. Thirdly, following the onset of the pandemic, an additional online virtual workshop was held in June 2020 that sought to 're-imagine' how the region could 'build forward better' after the pandemic co-organised with the enterprise development organisation, Mol Téic (Dingle Creativity and Innovation Hub). Finally, a session was held with Corca Dhuibhne Community Energy Group to discuss the energy scenarios developed, initial modelling results, and get feedback on the analysis.

Central to the approach was to collaborate with our local partners in *Dingle Peninsula 2030*. How these four groups inputted into the energy scenarios is outlined in Table 6.2. The partners were not just consulted on what the research plan should involve, but were rather actively part of delivering it. This is why it was important to identify what would be of most use to them.

6.4 Modelling Tool Used to Generate Energy Scenarios

The energy system was modelled using the LEAP (Low Emission Analysis Platform) software, which is a useful simulation tool for exploring energy scenarios [26]. A simulation tool like LEAP was favoured over an optimisation model as it relies on user input to define the energy system configuration as opposed to solving for a least-cost solution. This more easily facilitates the exploration of scenarios based on questions and proposals raised during stakeholder engagements [27].

Table 6.1 Summary of how the community was involved in energy planning

Stakeholder Groups	No	Description	Input into energy scenarios
Students from the area studying in University College Cork	13	Third level students in 3rd/4th year of study coming from a wide range of subject areas	Initial scoping exercise to capture key issues facing the area and get input on the proposed research process
Community meeting attendees	398	Residents from the area, generally representing older age groups	Highlighted key issues facing the area, priorities for future development, and additional measures to explore (such as active/public travel, renewable energy microgeneration)
Re-imagine workshop attendees	28	Selected to represent key groups: farming, tourism, hospitality, young people, ongoing sustainability initiatives, and social services	Framing of the build forward better scenarios and additional measures to be explored (such as active/public travel, renewable energy microgeneration)
Corca Dhuibhne Community Energy Group	8	Voluntary group exploring opportunities to establish an energy cooperative in the area	Feedback session held to discuss initial results and model inputs

Building the LEAP Model

LEAP is a simulation-based modelling tool, which means the user defines both the inputs and outputs. The majority of the work is performed externally in Excel, LEAP then functions as a useful tool to bring the various sectors (transport, housing, industry, etc.) together, explore variations on model parameters, perform calculations, and display results.

There are two central folders to be built in the model: key assumptions (Fig. 6.1 left-side) and demand (Fig. 6.1 right-side). Key assumptions are the input parameters such as the number of homes or cars, fuel

Table 6.2 Local and national partners the research team is actively collaborating with as part of *Dingle Peninsula 2030*

Stakeholder Groups	Description	Role in the project	Input into energy scenarios
North East West Kerry Development (NEWKD)	Non-profit community development organisation in the region	Providing link to existing community networks	Co-organising the community meetings held and ongoing discussion of energy system modelling scenarios/results
Dingle Creativity & Innovation Hub (Mol Téic)	Non-profit primarily focused on enterprise development in the region	Establishing the Sustainable Energy Community group and a number of other sustainability initiatives including Anaerobic Digestion feasibility study, farm ambassador programme	Co-organising the re-imagine Dingle Peninsula workshop, and ongoing discussion of energy system modelling scenarios/results
ESB Networks	National distribution system operator	Running a trial on the electrification of heat/transport in part of the region	High-level ambition for the electrification of heating and transport
Dingle Peninsula Sustainable Energy Community	Voluntary group formed out of Mol Téic to access grant support for energy planning and building energy improvements	Developed an energy masterplan, providing an estimate of energy-related CO_2 emissions in the region	Co-authored a report on energy-related CO_2 emissions in the region

shares and energy demand by technology, and emission factors for each fuel. These can be directly linked to an Excel sheet or inputted into the model. Future projections (for key parameters like the number homes or cars) is likewise flexible, drawing from Excel or using LEAP features to extrapolate from the base year to an end year value or growing at a particular rate. Within the demand folder, calculations are then setup using the key assumptions data. Figure 6.1 provides an example of the LEAP model structure, showing: branches (yellow folders), key assumptions (blue boxes), technologies (grey cogs) and environmental factors (dark grey cloud).

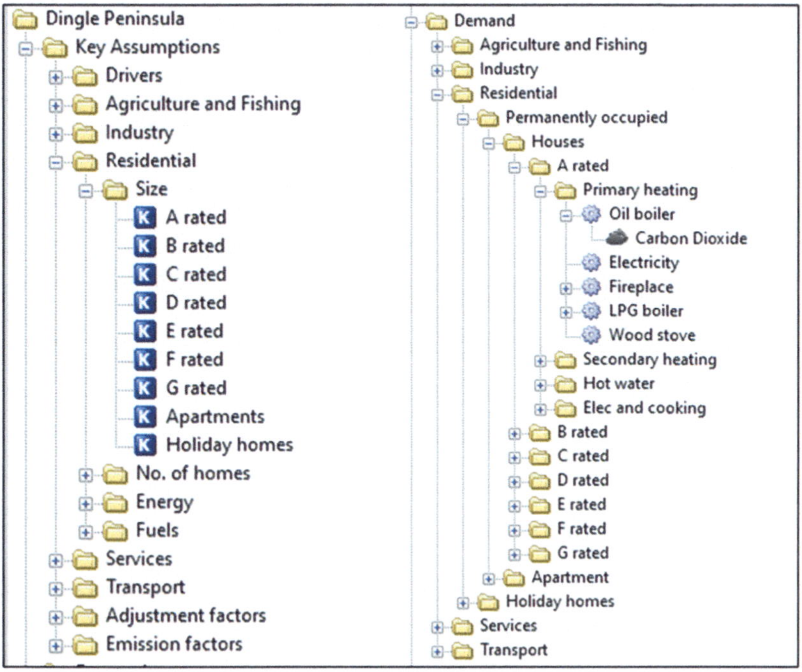

Fig. 6.1 Example of the LEAP model structure

The makeup of the branches in each sector differs based on the data available and intended analysis. For example, in the case of the residential sector, there are four key assumptions folders:

1. Size—Average size of homes by energy rating (as shown in Fig. 6.1).
2. No. of homes—Number of homes in each energy rating category (A, B, C, etc.).
3. Energy—Energy demand (kWh/m^2) for the different end uses (e.g. primary/secondary heating, lighting) in each energy rating category.
4. Fuels—Fuel share for each end use in each energy rating category, e.g. % share of A rated houses using oil for primary heating.

The calculations in the demand folder then pull these four parameters together. There are two elements to this: the activity level and final energy intensity. For example, in the residential sector:

$$Activity\ level = No.of\ Homes\ x\ fuel\ share \tag{6.1}$$

$$No.of\ A\ rated\ homes\ with\ oil\ boiler = 133\ x\ 12.6\% = 17 \tag{6.2}$$

$$Final\ energy\ intensity = Avg.m^2\ x\ Avg.kWh/m^2 \tag{6.3}$$

$$A\ rated\ home\ with\ oil\ boiler\ MWh\ per\ year = 224\,m^2\ x\ 23\,kWh/m^2 = 5.2\ MWh \tag{6.4}$$

$$Energy\ demand = Activity\ level\ x\ Final\ energy\ intensity \tag{6.5}$$

$$Subtotal\ for\ A\ rated\ homes\ with\ oil\ boiler = 17\,x\,5.2\ MWh = 87.5\ MWh \tag{6.6}$$

To determine energy-related CO_2 emissions, an emission factor (gCO_2/kWh) is then applied based on the technology, e.g. carbon dioxide under oil boiler in Fig. 6.1.

The capital cost of mitigation measures was not included. In several cases (e.g. electric vehicles) it would be possible to provide an estimate of the cost to the exchequer based on government incentive programmes. However, the actual of cost investing in the different mitigation measures (in particular building improvements), which is what would be of interest to community partners, is difficult to determine. There is very limited data available on the cost of retrofitting different houses or businesses, and estimates will vary significantly based on building characteristics. It was thus decided to omit investigation of capital costs.

The model has been made available for download from GitHub [28], and further explanation of the model can be found [29, 30].

Scenarios Overview

The implementation of scenarios in LEAP is likewise flexible. It can be done by rerouting the path for key assumptions to different cells in Excel, or within LEAP modifying parameters, and the in-built interpolation function. These three methods were each used in the Dingle Peninsula LEAP model:

1. The number of homes being retrofitted was varied by changing the projections in Excel to different scenarios.
2. The impact of COVID, seasonality of businesses, and distance travelled by cars was modelled by adding adjustment factors to the calculations in the demand folder. For example, based on a road traffic counter data on the number of journeys [31], private car travel was reduced by 24% in 2020/2021.
3. Growth in local solar PV was done using LEAP's interpolation function, which is based on start and end year values. Beginning at 0.102MWh and growing to 0.461MWh as follows: 'Interp (2020, 0.102, 2030, 0.461)'.

A wide range of measures were modelled in each sector to assess possible pathways (see Table 6.3). These were collated into five representative scenarios based on two potential reference scenarios; a return to pre-COVID-19 pandemic trends that would represent 'stagnation' versus a 'build forward better' scenario. Table 6.3 provides a brief overview of the scenarios, with particular focus on what happens with regard to residential and private cars, two important sectors identified in the baseline analysis (Sect. 6.5). The mitigation measures explored begin by looking at the impact of national policies (POL scenarios) [32], before implementing additional suggestions that emerged from the community engagements (COPROD scenarios).

A significant issue raised throughout the community meetings was the need to reduce the number of seasonally occupied holiday homes and provide houses for young families looking to settle in the area. In addition, during the 'Re-imagine Dingle Peninsula' workshop one of the central themes was how the region might capitalise on a move

Table 6.3 Overview of energy system scenarios modelled for the period 2010–2030

Name		Description
Reference	REF	Business as usual reference scenario with COVID-19 impact • COVID-19 impact over 2020/21 • 2022 onward return to business as usual • Population growth is significantly less than national projection for the period 2019–2030 at an annual average rate of 0.45% compared to 1.1% nationally
Current Policy Mid-Range Emissions (Reference Scenario Variant)	REFPOL1	Mid-range emission scenario for national 2030 targets • 30% of homes retrofitted to B rating, spread across housing stock. 80% of retrofitted homes get a heat pump • Private car Internal Combustion Engine (ICE) sales go to zero in 2030, battery electric vehicles make up 50% of sales by 2030 and 20% of car stock, with hybrids and plugin hybrids likewise accounting for 50% of sales in 2030 and 20% of car stock
Current Policy Low Emissions (Reference Scenario Variant)	REFPOL2	Low emission scenario for national 2030 targets • Retrofitting targets lowest rated homes (E, F and G), which account for roughly 30% of homes, eliminating this portion of the housing stock by 2030. 80% of retrofitted homes get a heat pump • ICE sales go to zero in 2030, battery electric vehicles grow to 100% of sales by 2030 and 40% of car stock

(continued)

Table 6.3 (continued)

Name		Description
Additional measures co-produced (Reference Scenario Variant)	REFCOPROD	Additional measures based on local dialogues • Reduced car dependence by 2030: o Improved public transport replaces 2% of private car kms o Active modes replace 5% of private car kms • Retrofitting targets lowest rated homes (E, F and G), which account for roughly 30% of homes, eliminating this portion of the housing stock by 2030. In addition, minimum post-works energy efficiency target raised from B rated to A rated homes. 80% of retrofitted homes get a heat pump • Solar PV (4MW farm and 4.5 MW rooftop) installations equivalent to 15% of electricity demand by 2030 • Small-scale wind turbines equivalent to 1.05 GWh installed by 2030 • Solid fuel is banned from 2027 onward, replaced with biomass, which over time comes from local native woodlands • Biomethane from anaerobic digestion replaces LPG and natural gas use in industry, services and homes
Build Forward Better	BFB	Build forward better after COVID-19 impact • COVID-19 impact over 2020/21 • 2022 onward build forward better o 10% of holiday homes remain permanently occupied from 2022 onward o By 2030, 25% of holiday homes have become permanently occupied o Population grows around 14% between 2019 and 2030, in line with County/National projections for that period
Current Policy Low Emissions (Build Forward Better Scenario Variant)	BFBPOL2	As above in POL2 under reference scenario

(continued)

Table 6.3 (continued)

Name		Description
Additional measures co-produced (Build Forward Better Scenario Variant)	BFBCOPROD	As above in COPROD under reference scenario but with additional: • Reduced car dependence by 2030: o Improved public transport replaces 5% of private car kms o Active modes replace 7.5% of private car kms o Work from home 2 days a week for employees that this would be suitable, around 40% of the workforce. Replaces 7% of private car kms • Retrofitting target left at B rating for lowest rated homes undergoing works. To reduce impact of growth, 520 C/D rated homes upgraded to A, equivalent to number going from holiday homes to permanently occupied

to remote work post-COVID. These are the narratives that the 'build forward better' scenarios attempt to represent. A key driver for understanding the scenarios is the population projections. In the reference scenarios, between 2019 and 2030 the population would grow at an average annual rate of 0.54%, around half the national average of around 1% [33], and similar to the previous 0.51% average annual growth seen between 2002 and 2016 [34]. During the period from 2002 to 2016, the region's population grew by 7% compared to a national average of 22% and county-wide growth of 12% (Ibid). In the 'build forward better' scenario with 25% of the holiday homes becoming permanently occupied, the population growth rate experienced in the region would be double that in the reference scenario at around 14% between 2019 and 2030, which would put the region back in line with county and national projections for that period. Effort was made to discuss key parameters like these with local partners but the feedback received was very limited.

6.5 Setting a Baseline for Energy, Emissions, and Community Wellbeing

A vital first step in the planning process is to establish the current makeup of the energy system [1]. This will highlight what fuels are being used (e.g. oil or electricity) and the services which they provide (e.g. heat or private travel). As already mentioned in Chapter 3, it is important to consider GHG emissions rather than just energy flows or balances. Although the future scenarios developed focus on the energy-related CO_2 emissions, a baseline emissions inventory was also prepared to provide this wider context. Another necessary baseline is to build an understanding of the key social trends in the area and what the community sees as the most pressing issues they face.

Baseline Energy Balance and Greenhouse Gas Emissions

Perhaps unsurprising given the rural and remote nature of the Dingle Peninsula, energy demand is dominated by private car travel and home heating (Fig. 6.2). There is a lack of industrial activity in the area, and although tourism is a large employer for the region, the level of activity in the services sector (measured in terms of employees) is still below the national average. Up to 45% of the workforce may be commuting outside the region for work [35]. Most of the energy comes from imported fossil fuels or the national electricity grid. This incurs a large cost for the community, it is estimated that the region spends ~ €37 million each year on imported fuels.

While the focus of this chapter, and indeed, most of the work on the Dingle Peninsula to date has been the energy system, it is important to consider the complete greenhouse gas emissions breakdown of an area (as noted in Chapter 3). The rural nature of the Dingle Peninsula means that energy-related CO_2 emissions represent just a third of the total GHG emissions. The expertise of the research team is in energy engineering and not agriculture or land use, therefore we maintain a focus on the necessary transformation to the energy system. However, the GHG baseline presented in Fig. 6.3 is a stark reminder of how difficult it will be for all of society to achieve climate neutrality.

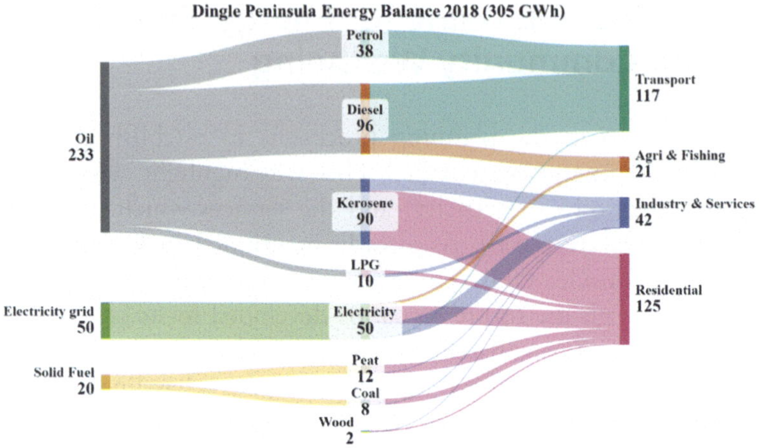

Fig. 6.2 Dingle Peninsula Energy Balance 2018

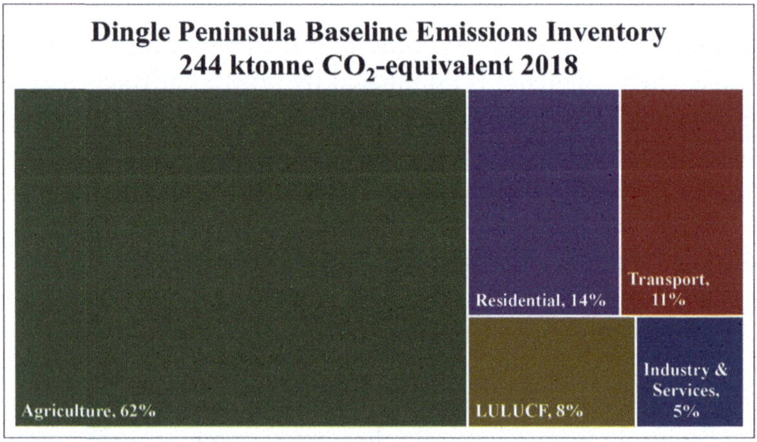

Fig. 6.3 Dingle Peninsula Baseline Emissions Inventory 2018

Community Engagement Findings

During both the initial student engagement and broader community consultation, a lot of the key concerns centred around the risk over-tourism posed to the area and limited opportunity for young people outside of this and the agriculture/fishing sectors. The students referred to this as the 'brain drain', those leaving the area to attend college rarely returned as the employment opportunities within Ireland's urban centres are much more diverse (Table 6.4). While all the students expressed a great pride in their home, they saw it as unlikely that they would settle there.

The first round of community meetings echoed the concerns the students had raised, as well as expanding on other issues such as care for the elderly, long-term viability of farming, and protecting the cultural heritage of the area (Table 6.5). This summary list of cross-cutting issues is useful to paint a picture of the region, but it is by no means exhaustive. There were many other issues discussed, some of which were unique to the experience of a particular community group. In general, while energy was often mentioned, it was not at the forefront of people's minds. For this community, getting rid of fossil fuels is just one part of the vision for a sustainable future, the top priorities are having a place for young families to settle and the elderly to be looked after.

Table 6.4 Summary of the student workshop discussion

Key themes	Questions emerging
The Dingle Peninsula is a small, tightly knit community	How can the Dingle Peninsula protect this strong sense of community into the future?
The Dingle Peninsula is a place of great natural beauty	How can Dingle ensure that increasing number of tourists and demand for infrastructure/development doesn't impact on the landscape?
'Brian drain'—Limited job opportunities outside of hospitality and agriculture means young people don't return after college	Need for better job opportunities—How can the Dingle Creativity and Innovation Hub support people working remotely?
Remoteness/Lack of infrastructure—no internet in many areas, no public transport and limited facilities for youth. Developers more interested in capitalising on tourism income rather than community needs	Could the future developments be better planned, considering the concerns of locals?

Table 6.5 Key concerns raised during the community meetings, 1st round November 2019

- Caring for the ageing population, access to shops, health services, etc
- Supporting community development and the work of community/ voluntary groups
- The expansion of facilities and amenities for young people
- Ensuring farming is competitive and that there are better linkages between farming and tourism
- Strengthening community relationships and resilience
- Protecting the magnificent culture, language, heritage and landscape of the area
- Reduce reliance on imported fossil fuels in favour of locally available renewables

It is hard to place the community development priorities in a hierarchy, as all are ultimately important to the long-term wellbeing of the region. However, there were four issues in particular that emerged as essential starting points: making housing available for young families looking to settle in the area, broadening the economic activity to ensure year-round income, improving connectivity through public transport or active travel infrastructure and ensuring there was appropriate housing as well as care for the ageing population (Table 6.6). Building on the community meetings, through the 'Re-Imagine Dingle Peninsula' workshop, five working groups were formed around key projects: a sustainable transport network, capitalising on a move to remote work post-COVID, developing the local food economy, mixed housing complexes for young and old families and green spaces.

The community meeting discussions provide two important considerations. Firstly, rural communities facing grave demographic, social, and economic challenges have very limited capacity for worrying about an existential threat like climate change. Secondly, many of the (un)sustainability issues are exacerbated by the fact that local development plans to do not adequately reflect the concerns and priorities of the community. In fact, in some cases the current challenges facing the Dingle Peninsula are direct consequences of national, regional (South West region of Co. Kerry and Cork) and county (Kerry County Council) policy objectives.

Table 6.6 Key priorities for future development, 2nd round of community meetings, February 2020

• Improved public transport, servicing all areas of the peninsula • Prioritise housing for full-time residents, not holiday homes • Development of mixed/sheltered housing complexes in town centres to cater for young and old • Develop year-round economic activity, not just the tourist season (May–Sept) • Address issues with sewage/wastewater treatment that are limiting ability to build new houses • Renovate the large number of vacant properties to newest energy standard • Support active modes of travel with improved paths/walkways as well as cycling infrastructure • Provide access to indoor and outdoor community areas • Promote the growth of native woodlands and rewilded areas • Installation of solar PV on all available rooftops: community centres, farm sheds and schools, etc

There was a lot of frustration at the limited access to housing driven by the National Planning Framework's priority of 'compact growth' [36], which restricts the building of any new one-off houses in rural areas. The local authority has a statutory obligation to implement this objective and thus extremely limited autonomy to address local housing concerns. However, there was a strong perception in the community meetings that the local authority is to blame for blocking housing proposals. The role of the local authority is poorly understood, and they are seen to be responsible as the ones who designate housing zones through forward planning and the processing (or rejecting) of planning applications. There were several accounts of young families seeking to build on their parent's land but being refused planning permission. This is preventing the intergenerational mixing that traditionally provided childcare and support for the elderly.

The purpose of the 'compact growth' objective is to address the connectivity challenges posed by Ireland's very dispersed population, which has resulted in a heavy dependence on private car travel. Much of rural Ireland suffers from forced car ownership due to the limited availability of public transport alternatives [37]. This issue was prominent in the community meeting discussions with a number of concerns raised; measures like a carbon tax will be overly punitive on rural people who have no option but to drive petrol or diesel cars, elderly people have difficulty accessing vital services, and businesses in town centres suffer because of the poor connectivity. The existing public transport service is extremely limited. For example, community 5 (see Fig. 6.4 below) had just a single service on a Friday connecting to the county capital, which is just beyond the peninsula's eastern boundary.

Key Socio-Economic Trends

The wellbeing indicators seen in Fig. 6.4 provide a useful overview of the issues identified during the community engagement process. It compares the eight communities that make up the case study region against the peninsula average (CD), county (KE), and national (IRE) values across key drivers of energy demand (car ownership, housing quality), demographic profile (share of young families, over 65s, etc.), and other socio-economic statistics (gross median income, broadband access). The values have been colour scaled to highlight those with the largest deviation in red from the national average which is green. It should be noted that the national or county averages are not necessarily desirable figures. However, a comparison in this manner nonetheless helps to highlight issues of concern by assessing how the area is faring relative to the rest of the county and country. The majority of values are based on the Census of Population of Ireland taken in 2016, as this was the most recent available at the time of investigation (Central Statistics Office, 2016a).

The two most peripheral communities (1 and 3) show the furthest deviation from the national average. This was reflected in discussions during the community meetings, in which there was frustration about those not on the development line (see N86 road in green on Fig. 6.4) between the mainland and the largest town in the area (Dingle, Community 2) being left behind. There is a compounding of issues in these communities. As the communities with the highest share of seasonally occupied holiday homes (up to 40% of houses), they thus have the lowest availability of homes, with only 50% of the housing stock being permanently occupied. The absence of young families is particularly striking in Community 3 where the share of young family households is -44% below the national average. In line with this, these communities have high shares of over 65s and people living alone while also having the worst quality housing (from an energy perspective). Unfortunately, it is not possible to determine the overlap between these sets of data, but it is likely quite high. This issue was a prominent topic in the community meetings, with great concern that the most vulnerable in the community may be experiencing energy poverty.

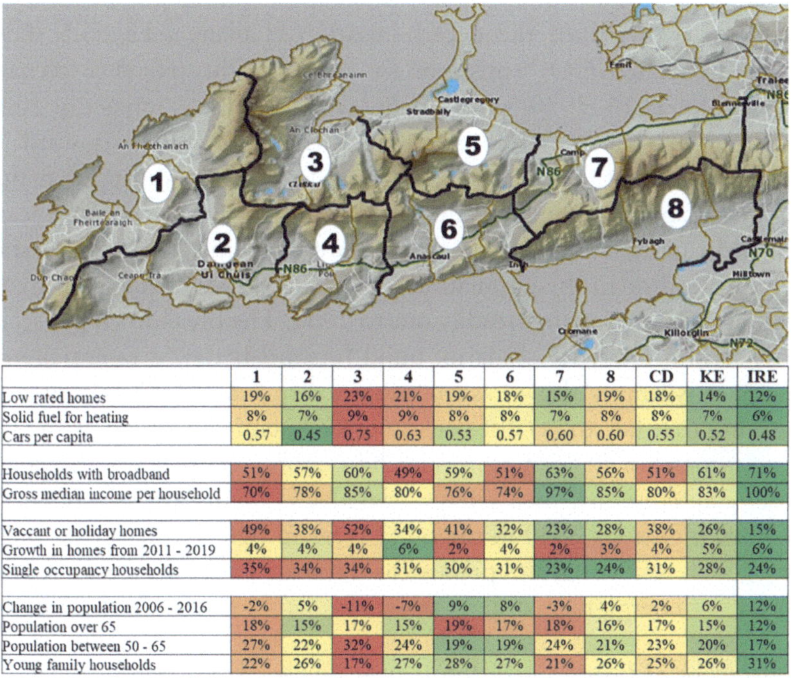

	1	2	3	4	5	6	7	8	CD	KE	IRE
Low rated homes	19%	16%	23%	21%	19%	18%	15%	19%	18%	14%	12%
Solid fuel for heating	8%	7%	9%	9%	8%	8%	7%	8%	8%	7%	6%
Cars per capita	0.57	0.45	0.75	0.63	0.53	0.57	0.60	0.60	0.55	0.52	0.48
Households with broadband	51%	57%	60%	49%	59%	51%	63%	56%	51%	61%	71%
Gross median income per household	70%	78%	85%	80%	76%	74%	97%	85%	80%	83%	100%
Vaccant or holiday homes	49%	38%	52%	34%	41%	32%	23%	28%	38%	26%	15%
Growth in homes from 2011 - 2019	4%	4%	4%	6%	2%	4%	2%	3%	4%	5%	6%
Single occupancy households	35%	34%	34%	31%	30%	31%	23%	24%	31%	28%	24%
Change in population 2006 - 2016	-2%	5%	-11%	-7%	9%	8%	-3%	4%	2%	6%	12%
Population over 65	18%	15%	17%	15%	19%	17%	18%	16%	17%	15%	12%
Population between 50 - 65	27%	22%	32%	24%	19%	19%	24%	21%	23%	20%	17%
Young family households	22%	26%	17%	27%	28%	27%	21%	26%	25%	26%	31%

Fig. 6.4 Wellbeing indicators identified in the 8 communities that make up the case study region compared to county and national figures (CSO data 2016), along with a map indicating the location of the communities

6.6 Scenarios for the Energy System in 2030

Drawing on a combination of inputs from community engagements, discussions with partners, and national climate policies a range of scenarios for 2030 were developed (Table 6.3). These begin with a reference scenario or business as usual before looking at the different interventions available in each sector to reduce CO_2 emissions.

Reference Energy Scenarios

To get an overview of the decarbonisation challenge, Fig. 6.5 shows the energy-related CO_2 emissions by sector in the reference scenario modelled for 2016–2030, which includes a COVID-19 impact in 2020/21 and then returns to pre-pandemic trends from 2022 onward. In response to the COVID-19 pandemic, and discussion during that time of a 're-imagined' Dingle Peninsula, there were two reference cases projected out to 2030: 1) a return to pre-pandemic trends with stagnation or decline in the population for most areas, 2) a change in trends with population growing steadily out to 2030. The breakdown per sector in the second reference scenario, build forward better, is essentially the same so it is simply shown as the projected increase in CO_2 emissions. Assuming that key drivers of per capita CO_2 emissions such as the rate of car ownership (0.55/person) and household occupancy (2.7 people/house) stay constant, then energy-related CO_2 emissions are projected to be 8% higher by 2030 in the build forward better scenario. As previously noted, residential and private car travel are key sectors, representing over half of energy-related CO_2 emissions.

Translation of Community Input into Energy Scenarios

The findings from the community engagement provided some elements of the scenarios (see in particular Table 6.7). How these inputs were represented in LEAP was discussed several times with the local partners and during a feedback session with the community energy group. This is an important part of the participatory process, as a means to verify the translation of participant inputs into the energy system model. For

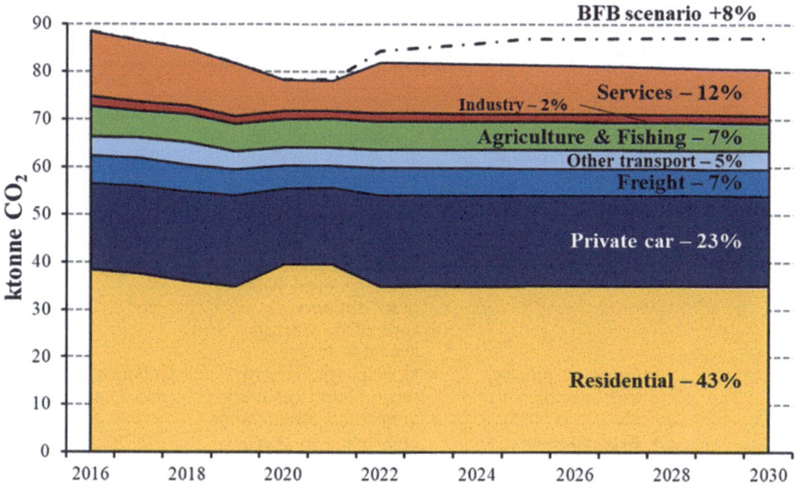

Fig. 6.5 Dingle Peninsula energy-related CO_2 emissions by sector in the reference scenarios 2016–2030

example, discussing what share of holiday homes became permanently occupied during the COVID-19 lockdown periods or what share of farm sheds should be covered in solar PV panels. However, it should be noted that a detailed discussion of these model parameters was generally not of interest so much of the decision-making remained in the hands of the research team. This is a critical challenge for participatory energy system modelling approaches [23]. We want to broaden the range of voices contributing to the process, but discussions with people from outside the energy field are difficult, as a basic understanding of the inner workings of the models is needed in order to effectively engage with them.

A key methodological challenge was making the link between the findings from community engagements and the energy scenarios. There were three common issues: areas of interest to the community neglected by national policy (e.g. rooftop solar PV, public transport, bioenergy), concerns of the community that couldn't be represented in the model (e.g. vacant homes, bus tours), and instances where there is tension between national policy and local concerns (e.g. compact growth, tourism).

Table 6.7 Comparison of national policy, local perspectives, and how they are represented in LEAP

Topic	National policy	Local perspectives	Representation in LEAP
Residential heating	Poorly defined retrofit and heat pump targets	• Concern about the amount of overlap between ~ 30% living alone, ~ 20% over 65 and ~ 20% lowest energy rated houses • Individual investments too high even with grant support • Lack of contractors available locally	Scenarios explored the impact of retrofitting lowest rated homes as opposed to an even share across dwelling types
Private travel	• Focus on private cars switching to electric vehicles • Active/public transport neglected, particularly in rural areas	• Most people rely on second-hand car market • Concerned about range and access to chargers in dispersed rural area • Lifecycle impact of EVs questioned • Buses a vital service for ageing population, and keeping small town businesses open • Active modes important for health, particularly in younger population • Better infrastructure can support more sustainable tourism model	Scenarios explored how increases in active and public transport may reduce private car demand
Renewable energy	• Focus on wind and solar • Heat and transport options other than electrification neglected • Favours large-scale developments • Limited supports available for small/ microgeneration	• Significant interest in microgeneration, and in particular rooftop solar PV • Tension between need to reduce reliance on imported fossil fuels and impact of large developments on landscape • Bioenergy of great interest: 1) native woodlands and 2) developing local circular economy	~ 10% local renewables by 2030 • Large uptake of solar PV on households, farm sheds, schools, etc • Native woodland crop developed for secondary home heating • Small-scale anaerobic digestion plant provides biomethane resource to replace LPG

(continued)

Table 6.7 (continued)

Topic	National policy	Local perspectives	Representation in LEAP
Holiday homes	No clear policies in place	• Some communities ghost towns in winter • Extremely limited long-term lets available	Within BFB scenarios: • Having been occupied during pandemic, 25% of holiday homes become permanently occupied by 2030 • No new builds become holiday homes
Vacant homes	Current policies are having limited impact	Frustration at high vacancy rates when young families can't find a home	None
Demographic imbalance	Compact growth places strong emphasis on urbanisation, essentially bans the building of new one-off houses	• Serious concerns about people's ability to live in remote areas • Large number (~30%) single occupancy households	Build forward better population increase based on current demographic makeup. However, uncertain what age groups this would involve
Tourism	Strongly promoted as a key economic sector, no clear policies to manage local impacts	• Needs to be more carefully managed, important opportunity but also a serious threat • Season limited to 6 months of the year, which is not viable for businesses • Large number (~50/day in summer peak) of daily bus tours coming from outside the area provide very little revenue locally while causing traffic issues and poor air quality	• Majority of businesses assumed closed in the winter • Within BFB scenarios it is assumed that the share staying open year-round grows • Tour buses not included in the transport sector, and air quality impact not considered

The fact that the scenarios informed by local interests achieve the lowest emissions savings is not so much reflective of the community's ambition as it is the misalignment between national policy and community interests. The measures based on community input are represented as additional to national policy, which at the time had neglected to include supports for microgeneration and had a narrow focus on electric vehicles in the transport sector. With respect to renewable energy,

Irish policy was strongly focused on large-scale developments up until 2021, and in particular wind energy, while generally neglecting the smaller/micro-scale technologies that were found to be of most interest to the community: rooftop solar PV, small-scale hydro, and bioenergy. In transport, public/active travel were seen to represent far more effective interventions than the electrification of private cars, which likewise was the central policy priority before 2021.

Other issues of importance to the community proved difficulty to represent in LEAP. The seasonality of businesses was possible to account for thanks to a previous building survey conducted [25]. However, the impact of tour buses (causing traffic and air pollution) was not included as they originate from outside the region. New builds within the model could be said to come from vacant homes as was suggested in the community meetings but this is of limited relevance to the energy scenarios. In addition, there are of course a wide range of issues across culture, heritage, social services, and wellbeing that are of vital importance to the community that are not considered in the energy system model.

Energy-Related CO_2 Emissions by Sector and Fuel in the Scenarios Modelled

To reduce CO_2 emissions, a range of measures were explored in each sector. These were primarily based on national targets for key issues like retrofitting homes, installing heat pumps, or electric vehicle uptake, as well as additional measures emerging from community engagements or local groups. Given the significance of transport and residential CO_2 emissions (as highlighted in Fig. 6.5), particular focus was given to those sectors.

Figure 6.6 gives a breakdown of the energy-related CO_2 in 2030 compared to a 2018 reference year for the seven scenarios outlined in Table 6.3 by sector in the top half and by fuel in the bottom half. The choice of 2018 as a reference year is based on the national target of a 51% reduction in greenhouse gas emissions by 2030 compared to 2018 levels or 7% per annum between 2021 and 2030 [38]. It should be noted that this is essentially double what the national target had been at the time of the community engagements and initial stages of the energy system model development, which was a 30% reduction by 2030 relative to 2005 or roughly 3.5% per annum between 2021 and 2030 [32].

In the residential sector, targeting low rated homes brings significant savings. The suggestion to increase the targeted energy efficiency from a B2 to an A rating was seen as important to avoid a lock-in of B rated homes unlikely to upgrade to A. However, it had little impact as with the installation of heat pumps the majority of the savings occur in electricity, which is expected to have very low-carbon intensity by 2030. Moving onto the transport sector, and in particular private car travel. The BFBCOPROD was significantly below BFBPOL2 in a large part due to efforts to reduce car dependence. With ~45% of workers commuting off the Peninsula for work, the potential reduction associated with working from home is quite significant.

Looking at the impact of each scenario on ktonne CO_2 by fuel source in 2030 highlights how key sources of emissions are dealt with or remain. Bioenergy in the form of biomass and biomethane can replace solid fuel and LPG/natural gas in the COPROD scenarios. While the CAP2 scenarios essentially eliminate solid fuel use in primary heating, the COPROD scenarios importantly also eliminate it in secondary heating. The petrol emissions also fall significantly in the higher electrified scenarios (CAP2 / COPROD) as small petrol engines or petrol hybrids are projected to be replaced rather than larger diesel engines. Diesel and heating oil continue to dominate energy supply and are the largest sources of associated CO_2 emissions in all scenarios. The heavy reliance on imported oil products remains a key concern, ranging from 60% to 75% of energy demand by 2030. This is in a large part due to the limited impact on diesel vehicles in private car travel and the lack of any interventions for heavy goods vehicles, tractors, or fishing boats.

Renewable Energy Contribution

During discussions with both representative organisations and residents, the ambition of being 'energy independent' sparked a lot of interest. However, turning that goal into reality is challenging. Despite widespread agreement that reducing energy imports should be a priority, there were two recurring difficulties that limited what renewable energy

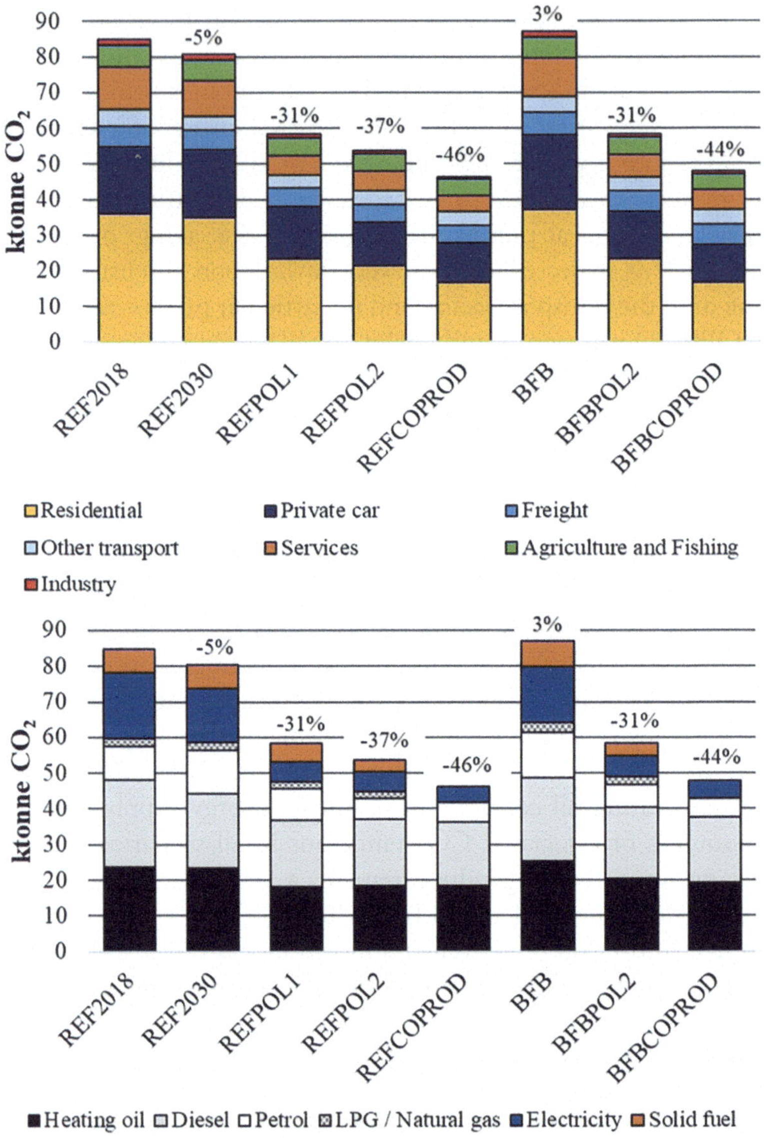

Fig. 6.6 Energy-related CO_2 emissions on the Dingle Peninsula by sector (top) and by fuel (bottom) for the end year 2030 in the eight scenarios modelled compared to reference year 2018

options the community wanted to pursue. Firstly, the local authority planning department does not consider large renewable installations appropriate for the area, since around a third of the region is protected as natural heritage sites (Special Protection Areas). Secondly, there was a preference in the community for small/micro-scale developments rather than larger projects. An overview of the renewable energy technologies considered is provided in Table 6.8.

Figure 6.7 provides an overview of the energy demand by fuel for each of the scenarios modelled, comparing the end year (2030) to the base year of 2018. The local renewable energy share is essentially zero, except in the COPROD scenarios when it is expected to be 11–12% of energy demand by 2030. Interestingly, the local renewable energy sources would account for just under 40% of renewable energy. The national blending of biofuels in transport fuels may contribute around 10 GWh of renewable energy, which is equivalent to the proposed local anaerobic digestion plant or Solar PV developments. This was seen as an effective action in the community meeting as it doesn't require any individual action or investment to switch to new technologies. However, it could be criticised as failing to address the problems associated with a car dependent transport system.

Table 6.8 Overview of the renewable energy sources considered for the case study region

Source	Identified resource	Chosen annual output
Solar PV	4 MW solar farm	3.9 GWh
	4.5 MW rooftop Solar PV	4.3 GWh
Wind energy	3 × 4 MW turbines	29.4 GWh
	80 × 5 kW pole-mounted microturbines	1.05 GWh
Small-scale hydro	530 kW from 6 turbines ranging from 30 to 180 kW	2.4 GWh
Bioenergy	Biomass—1000 ha native forestry	8.2 GWh
	Anaerobic digestion	10 GWh

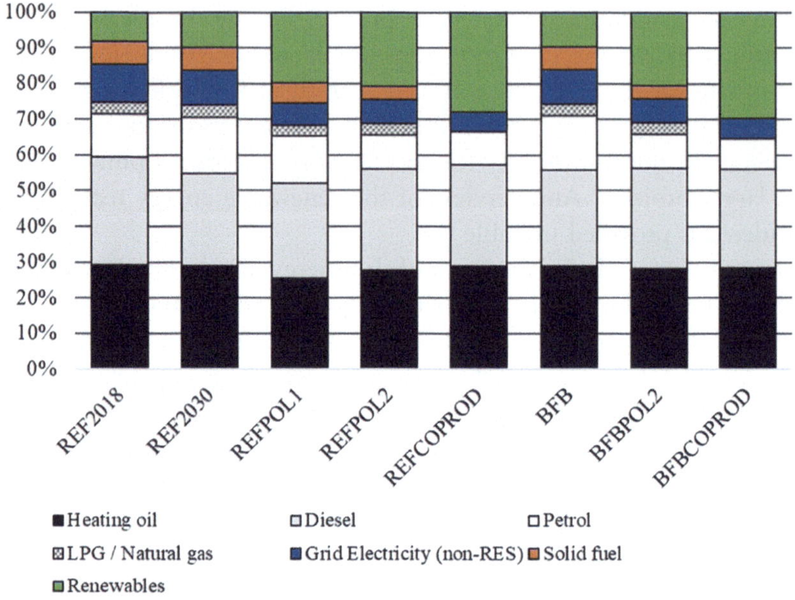

Fig. 6.7 Energy supply by fuel on the Dingle Peninsula for the end year (2030) of the seven scenarios modelled compared to base year (2018)

6.7 Discussion

What Does the Energy System Model Tell Us About Pathways for the Dingle Peninsula?

The energy system modelling exercise itself provides important insights on the challenge decarbonisation poses for rural areas. Firstly, the scale of the challenge is laid bare when dealing with such a small rural area. Secondly, while high-level national ambition like retrofitting 450,000 homes by 2030 is important, the pathway to actually delivering it requires more careful consideration. Thirdly, there is a significant gap between the community objective of energy independence and preferred renewable energy options.

Home heating and private travel are key issues. For home heating, the results demonstrate the need to consider which houses are being retrofitted and eliminating solid fuel (i.e. peat and coal). The reduction in emissions within the residential sector within REFPOL1 whereby the retrofitting is performed evenly across the housing stock is 36%, but in REFPOL2 by targeting the lowest rated houses a 41% reduction is achieved. Having addressed the lowest rated homes, one key source of emissions left is the roughly 5ktonne CO_2 (~25% of CO_2 emissions in the sector in 2030 under REFPOL2) associated with secondary heating (i.e. fireplaces and stoves). At present even in the highest rated homes (A/B), solid fuel is very common in secondary heating at 55–65% of fuel share. With regard to private travel, the results show that electric vehicles will not be sufficient, and thus more transformative interventions are needed to support active modes and public transport, which until recently were neglected by national policy.

What is most striking about the results is that the highly ambitious COPROD scenarios modelled for the case study region results in a 44–46% reduction in 2030 relative to 2018 (Fig. 6.6). Far short of what is needed to achieve a 51% reduction across all sectors, since energy will need to do more in order to accommodate a less significant reduction in non-energy emissions from agriculture. These results clearly highlight that existing policy measures do not match emission reduction targets, and that is before even considering the broader societal issues facing the community. It should be noted that the scenarios modelled were based on the National Climate Action Plan 2019 [32]. This plan is updated each year, but there has been very little change in the key targets for retrofitting, heat pumps and electric vehicles.

Just as the participatory process shines light on the limitations of the energy system model, the technical analysis likewise points to contradictions within the community objectives. The local renewable energy contribution is expected to be limited at only 11–12% of energy demand in 2030 in the most ambitious scenarios, which is a long way from the aspiration of energy independence. If that goal is to become a reality, difficult decisions need to be made on the placement of much larger energy developments than what is currently deemed acceptable, as well as exploring what actions can be taken to reduce demand. Another key tension is the desire for significant growth in population while at the same time reducing emissions.

What Did the Transdisciplinary Approach Tell Us About the Energy System Model?

The most valuable learning is the differences between the *messiness* of the real-world compared to the simplicity of the energy system model world. The community engagement process was intentionally open to the broad range of concerns people have, which posed a serious methodological challenge but also opened up some critical conversations about the future of the region. In this case study, taking a purely technical perspective would omit consideration of the serious societal capacity issues. This context is crucial to identify barriers and develop an understanding of how to implement the necessary measures. For example, looking at housing, the modelling results clearly show a need to address the poorest quality houses. However, it is only by looking at the broader picture and discussing it with the community that we can understand who lives in these homes and what sort of policy interventions are needed.

On from this, a key element is the identification of co-benefits. When energy/climate policies are aligned with community development needs, then climate action presents a great opportunity to address some of the inequities in society such as the growing gap between rural and urban populations seen here. The provision of public transport is an effective way to reduce CO_2 emissions, while also providing an essential social service. Active modes offer health benefits, improved air quality and, important in this case study region, infrastructure for sustainable tourism. Similarly, dealing with the highest emitting homes and ensuring the ageing population has appropriate housing can be closely aligned goals.

The benefits of this alignment between community development and climate action are clear but actually realising them will require more careful policy planning. The areas of interest to the community (solar PV, public/active transport, bioenergy) have been neglected by national policy until recently. This means opportunities for community participation in climate action are being missed. On from this, there is the delivery of services. With regard to housing, given the likely overlap between poorly insulated homes reliant on solid fuel and an elderly population living alone, managing the necessary disruption to improve the energy performance of these homes will require careful coordination.

6.8 Conclusion

There are a wide range of considerations and initiatives emerging out of the Dingle Peninsula that cannot be adequately represented in an energy system model. Moreover, it must be recognised that had the original research plan been followed and discussions limited to what was relevant to the energy system model, much of the broader societal context would have been omitted. It is important to appreciate the complexity and messiness of the real world in order to develop effective climate and energy policy interventions. Firstly, there is the question of current (un)sustainability challenges across environmental and social domains, and what is driving them. The socio-economic and demographic challenges facing the Dingle Peninsula in housing, employment, and an ageing population means the capacity to respond to the climate crisis is limited. This calls for a broadening of our understanding of a desirable future beyond just the technical changes associated with decarbonising our energy system.

There is also the question of who can pay for the energy transition. Ireland's decarbonisation strategy at present relies heavily on individual households making the investment in an electric vehicle or home energy improvements. This is excluding a large portion of the population who do not have the resources and capacity to access the grant supports, which poses a serious equity and distributional justice issue. As Social Justice Ireland highlight "*these subsidies are functioning as wealth transfers to those households on higher incomes while the costs (for example carbon taxes) are regressively socialised among all users*" [39]. Under current policy, it seems likely that the gap between rural and urban populations demonstrated will worsen with the households most in need of energy improvements (offering the most significant CO_2 savings) unable to make use of grant supports. This will leave rural communities feeling excluded and thus rightly frustrated, which will make issues like the carbon tax and placement of energy infrastructure increasingly divisive. This poses a serious risk to decarbonisation efforts.

More open and transparent dialogue is needed to explore tensions between local and national priorities in order to develop well-informed policy. This would necessarily be a two-way process as co-benefits are not always guaranteed. There will be instances where higher-level policy decisions can be aligned with local needs but there will also be trade-offs and difficult issues that need to be resolved. In line with policymakers being more informed of 'on-the-ground' issues, people and groups must likewise learn to appreciate the key constraints that policymakers are trying to juggle. It is only from such a point of mutual understanding of both the problem and available solutions that an agreement on the best path forward can be reached.

References

1. C. McGookin, B.Ó. Gallachóir, E. Byrne, An innovative approach for estimating energy demand and supply to inform local energy transitions. Energy **229**, 120731 (2021)
2. S. Rivas, G. Melica, V. Palermo, J.-F. Dallemand, Guidebook "How to develop a Sustainable Energy and Climate Action Plan (SECAP)". Part 1, The SECAP process, step-by-step towards low carbon and climate resilient cities by 2030 (2018). https://op.europa.eu/en/publication-detail/-/publication/338a9918-f132-11e8-9982-01aa75ed71a1/language-en. Accessed 22 June 2024
3. European Commission, Home Why a Covenant of Mayors? (2019). https://eu-mayors.ec.europa.eu/en/about. Accessed 22 May 2024
4. European Commission, Home Key figures (2024). https://eu-mayors.ec.europa.eu/en/key_figures. Accessed 22 May 2024
5. C. McGookin, B.Ó. Gallachóir, E. Byrne, Participatory methods in energy system modelling and planning—a review. Renew. Sustain. Energy Rev. **151**, 111504 (2021)
6. D. Reckien, K. Eisenack, Climate change gaming on board and screen: a review. Simul. Gaming **44**, 253–271 (2013)

7. S. Flood, N.A. Cradock-Henry, P. Blackett, P. Edwards, Adaptive and interactive climate futures: systematic review of 'serious games' for engagement and decision-making. Environ. Res. Lett. **13**, 063005 (2018)
8. J. Sterman, T. Franck, T. Fiddaman, A. Jones, S. McCauley, P. Rice, E. Sawin, L. Siegel, J.N. Rooney-Varga, WORLD CLIMATE: a role-play simulation of climate negotiations. Simul. Gaming **46**, 348–382 (2015). https://doi.org/10.1177/1046878113514935
9. M. Thomas, T. Partridge, N. Pidgeon, B.H. Harthorn, C. Demski, A. Hasell, Using role play to explore energy perceptions in the United States and United Kingdom. Energy Res. Soc. Sci. **45**, 363–373 (2018). https://doi.org/10.1016/j.erss.2018.06.026
10. A. Krzywoszynska, A. Buckley, H. Birch, M. Watson, P. Chiles, J. Mawyin, H. Holmes, N. Gregson, Co-producing energy futures: impacts of participatory modelling. Build. Res. Inform. **44**, 804–815 (2016). https://doi.org/10.1080/09613218.2016.1211838
11. P. Nabielek, H. Dumke, K. Weninger, Balanced renewable energy scenarios: a method for making spatial decisions despite insufficient data, illustrated by a case study of the Vorderland-Feldkirch Region, Vorarlberg, Austria. Energy Sustain. Soc. **8**, 5 (2018)
12. F. Steinberger, T. Minder, E. Trutnevyte, Efficiency versus equity in spatial siting of electricity generation: citizen preferences in a serious board game in Switzerland. Energies **13**, 4961 (2020)
13. J. Flacke, C. De Boer, An interactive planning support tool for addressing social acceptance of renewable energy projects in the Netherlands. ISPRS Int. J. Geo-Inf. **6** (2017). https://doi.org/10.3390/ijgi6100313
14. M. Uwasu, Y. Kishita, K. Hara, Y. Nomaguchi, Citizen-participatory scenario design methodology with future design approach: a case study of visioning of a low-carbon society in Suita City, Japan. Sustainability **12**, 4746 (2020)
15. D.L. Bessette, J. Arvai, V. Campbell-Arvai, Decision support framework for developing regional energy strategies. Environ. Sci. Technol. **48**, 1401–1408 (2014)
16. L.A. Mayer, W. Bruine de Bruin, M.G. Morgan, Informed public choices for low-carbon electricity portfolios using a computer decision tool. Environ. Sci. Technol. **48**, 3640–3648 (2014)
17. G. Xexakis, R. Hansmann, S.P. Volken, E. Trutnevyte, Models on the wrong track: model-based electricity supply scenarios in Switzerland are not aligned with the perspectives of energy experts and the public. Renew.

Sustain. Energy Rev. **134**, 110297 (2020). https://doi.org/10.1016/j.rser.2020.110297
18. S.P. Volken, G. Xexakis, E. Trutnevyte, Perspectives of informed citizen panel on low-carbon electricity portfolios in Switzerland and longer-term evaluation of informational materials. Environ. Sci. Technol. **52**, 11478–11489 (2018)
19. C. Demski, A. Spence, N. Pidgeon, Effects of exemplar scenarios on public preferences for energy futures using the my2050 scenario-building tool. Nat. Energy **2**, 17027 (2017). https://doi.org/10.1038/nenergy.2017.27
20. T. Schinko, S. Bohm, N. Komendantova, E.M. Jamea, M. Blohm, Morocco's sustainable energy transition and the role of financing costs: a participatory electricity system modeling approach. Energy Sustain. Soc. **9**, 1 (2019). https://doi.org/10.1186/s13705-018-0186-8
21. B. Droste-Franke, M. Voge, A. Kanngießer, Achieving transparency and robustness of regional energy scenarios by using morphological fields in inter-and transdisciplinary project groups. Energ. Strat. Rev. **27**, 100430 (2020)
22. K. Eisenack, D. Reckien, Climate change and simulation/gaming (2013)
23. C. McGookin, D. Süsser, G. Xexakis, E. Trutnevyte, W. McDowall, A. Nikas, K. Koasidis, S. Few, P.D. Andersen, C. Demski, Advancing participatory energy systems modelling. Energy Strateg. Rev. **52**, 101319 (2024)
24. D.J. Lang, A. Wiek, M. Bergmann, M. Stauffacher, P. Martens, P. Moll, M. Swilling, C.J. Thomas, Transdisciplinary research in sustainability science: practice, principles, and challenges. Sustain. Sci. **7**, 25–43 (2012)
25. C. McGookin, D. de Bhailís, B. Tuohy, B.Ó. Gallachóir, E. Byrne, Dingle Peninsula Energy Master Plan (2020). https://dinglepeninsula2030.com/projects/energy/energy-master-plan/. Accessed 23 May 2024
26. C.G. Heaps, *Long-Range Energy Alternatives Planning (LEAP) System* (Stockholm Environment Institute, Somerville, MA, 2012)
27. M. Nilsson, A. Dzebo, G. Savvidou, K. Axelsson, A bridging framework for studying transition pathways—from systems models to local action in the Swedish heating domain. Technol. Forecast. Soc. Chang. **151**, 119260 (2020)
28. C. McGookin, MaREI-Energy Policy & Modelling Group GitHub Repository/LEAP_Dingle_Peninsula (2021). https://github.com/MaREI-EPMG/LEAP_Dingle_Peninsula. Accessed 20 Aug 2024
29. C. McGookin, T. Mac Uidhir, B.Ó. Gallachóir, E. Byrne, Doing things differently: bridging community concerns and energy system modelling

with a transdisciplinary approach in rural Ireland. Energy Res. Soc. Sci. **89**, 102658 (2022)
30. C. McGookin, Developing participatory methods in energy system modelling and planning. PhD Thesis, University College Cork (2022)
31. Transport Infrastructure Ireland, Traffic Count Data (2022). https://www.nratrafficdata.ie/c2/gmapbasic.asp?sgid=ZvyVmXU8jBt9PJE$c7UXt6. Accessed 23 May 2024
32. Department of Communications, Climate Action & the Environment, Climate Action Plan (2019). https://www.dccae.gov.ie/en-ie/climate-action/publications/Documents/16/Climate_Action_Plan_2019.pdf. Accessed 22 May 2024
33. Central Statistics Office, Population and Labour Force Projections 2017–2051 (2017). https://www.cso.ie/en/releasesandpublications/ep/p-plfp/populationandlabourforceprojections2017-2051/populationprojectionsresults/. Accessed 22 May 2024
34. Central Statistics Office, Census 2016 small area population statistics (2016). https://www.cso.ie/en/census/census2016reports/census2016smallareapopulationstatistics/. Accessed 23 May 2024
35. C. McGookin, B.Ó Caoimh, S.O'Hara, B.Ó Gallachóir, E. Byrne, Corca Dhuibhne/Dingle Peninsula 2030 learning briefs: evidence-based community planning on Corca Dhuibhne (2021). https://www.marei.ie/wp-content/uploads/2021/07/Evidence-based-planning-Corca-Dhuibhne_V2.pdf. Accessed 22 May 2024
36. Government of Ireland, Project Ireland 2040: national development plan 2018—2027 (2018). https://assets.gov.ie/37937/12baa8fe0dcb43a78122fb316dc51277.pdf. Accessed 23 May 2024
37. P. Carroll, R. Benevenuto, B. Caulfield, Identifying hotspots of transport disadvantage and car dependency in rural Ireland. Transp. Policy **101**, 46–56 (2021). https://doi.org/10.1016/j.tranpol.2020.11.004.
38. Department of the Environment, Climate and Communications, Climate Action Plan 2021: securing our future (2021). https://www.gov.ie/en/publication/6223e-climate-action-plan-2021/. Accessed 23 May 2024
39. Social Justice Ireland, Energy Poverty and a Just Transition (2021). https://www.socialjustice.ie/content/policy-issues/energy-poverty-and-just-transition. Accessed 23 May 2024

7

The Diffusion of Sustainability and Scaling Deep

Evan Boyle

7.1 Introduction

More action-orientated research is needed to support sustainability goals, aligning with our discussion as outlined in Chapter 2, however, there have been limited examples of how to monitor or measure the contribution of such research projects. Understanding the impact of our project posed a serious methodological challenge due to its emergent and evolving nature (See: Chapter 9). Furthermore, the exact contribution of different interventions or actors is hard to classify. There is a clear need to develop new metrics to understand what the impact of these embedded research projects may be. However, ways to understand this might not fit with traditional KPIs and project reporting structures.

Considering this challenge, we used the sustainable development goals (SDGs)to map the projects that have emerged in the case study area of *Dingle Peninsula 2030*. The unique approach leads to the concept of the *diffusion of sustainability* to understand impact. It is not a universal method to measure impact, but helps us point to a wide breadth of novel initiatives beyond the original objectives of the project. The core focus of the *Dingle Peninsula 2030* partnership was on projects related to energy

C. McGookin and E. Boyle, *Engaging Communities in Climate Action*, Energy, Climate and the Environment, https://doi.org/10.1007/978-3-031-93498-8_7

and climate. However, there are also successes in education (SDG 4), decent work (SDG 8) and innovation (SDG 9), which reflect the broader concerns of the community. This points to the importance of such place-based approaches to align vital sustainability objectives with community goals, and the holistic and interlinked nature of the SDGs when enacted.

Case Study Background

Dingle Peninsula 2030 emerged to facilitate and support decarbonisation initiatives. Its core aim was to bring sustainable employment to the area. The project used a novel governance structure by forming a collaborative committee. This committee included representatives from a local not-for-profit focused on enterprise (Mol Téic) and community development (NEWKD), as well as Ireland's national electricity distribution system operator (ESB Networks) and our research institute (MaREI). As a group we were grounded within the local community, whilst also having capacity at the national level. From a research perspective, we have studied *Dingle Peninsula 2030*, looking at social network analysis [1], the engaged research approach deployed, the collaborative governance structure [2], and participatory methods for energy system modelling [3].

Dingle Peninsula 2030 acts as a community-based, transdisciplinary collaboration that has had an impact both at a local community level and on a broader national and international stage. It offers valuable learnings on the real-world transformation processes to guide efforts elsewhere [4]. The project gathered national media coverage, was designated as a living laboratory by the United Nations and sent a delegation to COP-26 to discuss the role of community-based initiatives for climate action. Regarding the diffusion of sustainability, we focus the investigation here on the period from January 2018 until June 2021.

Diffusion of Sustainability

Niche social and technical innovations are emerging practices that differ radically from the prevailing socio-technical system and regime but can gain a foothold in particular applications, geographic areas, or with the help of targeted policy support. Community contexts are spaces for experimentation where new niche social, technical and governance ideas can be tested and take hold (as seen in Chapter 3). The global challenge of climate change impacts people locally and due to this, the local level can often bring about novel solutions that operate without the weight of loftier policies, strategies, and bureaucracies. While niche- innovations often act within specific and unique community contexts, they can be extrapolated from to provide wider learnings. Although community action can assist in informing policy developments, there is a need to create mechanisms that can represent instances of success. Here, the diffusion of sustainability is offered as a framework to help with this process of representation.

Recent work on the diffusion of transformative innovations for sustainability has highlighted several different categorisations for situating diffusion [5]. From this, a typology of development mechanisms for transformative innovation has been formatted. A number of these have relevance to the case study context presented in this paper [5].

- **Growing**, whereby an initiative quantitatively grows by attracting more participants and funding, can be facilitated by increased social visibility, professionalisation and communication capacities.
- **Replication**, whereby ideas or practices are translated into new contexts, can often occur due to ideas percolating in the media or communication channels.
- **Partnering**, whereby initiatives share resources and capacities, takes place when collaboration is mutually beneficial. These three typologies have acted to frame our understanding of the diffusion of sustainability in the context of *Dingle Peninsula 2030,* along with established social theory related to diffusion.

Diffusion can occur through three mechanisms: internal influence, external influence and mixed influence. At a basic level, internal influence in the diffusion of innovation refers to the spread of new ideas and practices through interpersonal communication among individuals [6, 7]. External influence refers to the diffusion of ideas and innovations through sources that are not related to personal relationships. The media plays a significant role in this external influence, along with triggers from the private sector and developments in policy. Additionally, mixed influence involves a combination of both internal and external principles, where the broader environment and interpersonal relationships collectively affect the adoption of innovations.

In recent years, there has been a notable increase in research focused on the spread of social innovations aimed at achieving sustainability. Studies examining social innovations in isolated rural areas, such as our own case on the Dingle Peninsula, have highlighted the significance of both local and external actors in facilitating change. While external dimensions play a role, these studies emphasise that local knowledge and community activities are crucial to the transformation processes in rural settings [8]. The regional level is understood within the literature to be an important context for the diffusion of sustainability to other locations [9]. *Dingle Peninsula 2030* has contributed to raising awareness of regional innovations for sustainability, and the diffusion mechanisms at play.

Diffusion is a process through which an innovation can be communicated using certain mediums or channels over time to members of a social group or system [10]. Diffusion can be understood as a "*cascading mechanism that leads to cumulative adoption of behaviours by some individuals even while their social position, or the resources associated with them, changes only trivially or remains unaltered*" ([11], p.68).

Borrowing from Clough et al. sustainability can be defined as:

> a process that helps create a vibrant economy and a high quality of life, while respecting the need to sustain natural resources and protect the environment. It expresses the principle that future generations should live in a world that the present generation has enjoyed but not diminished ([12], p.30).

Our definition for the diffusion of sustainability is understood to be aligned with the definitions of both Palloni and Clough et al., whereby initiatives, projects or actions which support the creation of a vibrant economy and high quality of life, while respecting natural resources and the environment, are cumulatively developed or adopted. This definition builds upon social innovations approaches to transitions whereby the centre of change under investigation is changing social practice rather than technological approaches to innovation [13]. Social innovations as a field of investigation has been rapidly growing within the academic literature in recent times [14].

The sustainable development goals (SDGs) can be used to capture social innovations, with nine socially orientated goals represented within the framework (SDG 1, 2, 3, 4, 5, 8, 10, 11, 12) [14]. Our own use of the "diffusion of sustainability" as a heuristic devise for recording the emergence of novel innovations in sustainability transitions is built upon the SDGs. Using an SDG framing can help understand not only social innovations but can also look towards technological/infrastructural innovations (SDG 7, 9), environmental/ ecological innovations (SDG 6, 13, 14, 15) or institutional/ governance innovations (SDG 16, 17). The sustainable development goals have been criticised for being hard to quantify, implement and monitor [15]. Some progress has been made with relation to these challenges [16, 17], yet they remain more progressive in their aspiration then their impact. *Dingle Peninsula 2030* is used to outline the *diffusion of sustainability* regionally, providing a foundation for measuring impact.

Developing a Process for Investigating Diffusion

This research investigates diffusion through qualitative methods, including exploratory interviews, ethnographic attendance at events and secondary research. The diffusions were cross-referenced with the primary project manager for Dingle Hub/Mol Téic. The goal was to outline emergent projects and initiatives stemming from *Dingle Peninsula 2030,* focusing on those involving the four core partners between January 2018 and June 2021. Additionally, I mapped spin-off projects led by local actors who participated in *Dingle Peninsula 2030* events and moved on to promoting their own sustainability initiatives.

A key challenge associated with this task is the fact it may takes years for transformations to take place. In the long term, a more in-depth data-centric approach to diffusion research would be insightful. The development of retrospective historical analysis of social diffusion is an area in which new mechanisms and methods are being developed, informed by contemporary network concepts and models [18]. Such approaches would provide useful means of investigation if one was to set about interpreting the diffusion of sustainability in Dingle at a later date. Here, however, we have sought to outline an approach which can chart and categorise diffusion in real-time, as part of a rapidly evolving project.

Various diffusion models have been developed in sociology for different social situations. However, their suitability can be limited due to the need to account for factors like socio-economic status, preferences, decision-making processes and individual constraints. As has been noted, "*the conditions for identification of a diffusion process from observables are fairly hard to meet, much harder than what is normally implied in traditional applications of diffusion theories and models to sociological and demographic analysis*" ([11], p.67).

Studying the diffusion of sustainability, in relation to environmental issues, is a well-developed field within tourism [19–21]. For our purposes, however, an investigation into the diffusion of sustainability in relation to regional transition projects is missing from the literature. It builds upon the three typologies related to growing, replication and partnering outlined previously [5]. The potential for sustainability to spread within and between communities, as well as through national initiatives, is crucial for achieving a low-carbon society. While less quantifiable than switching from petrol to electric vehicles or retrofitting homes, the diffusion of sustainability as an idea can significantly influence the development of community projects across various sectors.

In relation to *Dingle Peninsula 2030*, varying diffusions raise issues concerning the theories, assumptions and constraints needed in the model. Factors influencing technology diffusion differ significantly from those affecting climate education. Increased attention to decarbonisation and sustainability necessitates those external influences on the diffusion process be considered. It would be interesting to revisit this exercise in 10 years or even a longer timeframe to see how it has developed. A broader time frame for a comprehensive study of diffusion is essential. Key diffusion processes are summarised below (Fig. 7.1), using data from exploratory interviews, events and secondary research, and associated with relevant SDGs.

Fig. 7.1 Visual representation of the diffusion of sustainability as part of *Dingle Peninsula 2030*

Outline of Diffusions of Sustainability from Dingle Peninsula 2030

In this section, the diffusions are highlighted through a visual aid and a table providing a contextual outline. How the project and the emergent initiatives within it develop in the coming years will only become known with time. The information contained here represents up to 2021.

Table 7.1 outlines the SDGs linked to diffusions to date. Two of the goals were not listed as they are cross-cutting objectives of the overall project: SDG 11 (sustainable cities and communities) and 17 (partnership for the goals). The core focus of the initiative in energy (SDG 7) and climate (SDG 13) is clear, but there are also projects in education (SDG 4), decent work (SDG 8) and innovation (SDG 9).

Table 7.1 Diffusions of sustainability mapped against the sustainable development goals [27]

Diffusion	Sector(s)	Linked Sustainable Development Goal
1. Decarbonisation Zone	Region chosen as county's designated area for trialling climate mitigations measures	3, 7, 9, 12, 13
2. Corca Dhuibhne Community Energy	Group formed to work on energy-related initiatives for community benefit on the peninsula	7, 12, 13
3. *Dingle Peninsula 2030* SEC	Created as part of the national Sustainable Energy Community network with a focus on residential retrofits and energy production	7, 12, 13
4. West Kerry Dairy Farm SEC	A group of 100 farmers investigating solar production in the agricultural sector	6, 7, 8, 12, 13, 15
5. Dún an Óir Residents Association	55 members of residents' association exploring collaborative energy efficiency upgrades	7, 13
6. Energy Clinics	Facilitated through Dingle Hub, providing online advice on low-carbon technologies for home upgrades	7, 13
7. Solar Beo Expansion	Company established to provide clean energy upgrades	7, 8, 9, 13
8. KETB Community Energy Mentor Course	Course created providing training to 10 local people to act as energy mentors	4, 7, 13
9. KETB Regenerative Farming Course	Course developed on biological farming and regenerative practices	4, 13, 15

(continued)

Table 7.1 (continued)

Diffusion	Sector(s)	Linked Sustainable Development Goal
10. Anaerobic Digestion Feasibility Study	A group of stakeholders in the agricultural sector have formed a steering group to investigate the potential for Anaerobic Digestion in the area	7, 12, 13, 15
11. EU SKIN Project	An EU funded project investigating localised supply chains in the hospitality and agricultural sectors	3, 8, 15
12. EU Ploutos Project	An EU funded project investigating the use of sensors and data for environmental parameters in the agricultural sector	3, 8, 9, 15
13. EU Smart Rural Areas in the 21st Century	Selected as representative in Smart Rural 21 initiative, to promote and inspire villages/rural towns to develop and implement smart village approaches and strategies	3, 7
14. Creative Climate Action Project	Nationally funded project with local farmers to creatively look at ways in which farmers on the peninsula can diversify to address climate change	12, 13, 15
15. Sustainable Travel Pilot Project	Steering group established to investigate numerous local transport initiatives	3, 9, 12, 13

(continued)

Table 7.1 (continued)

Diffusion	Sector(s)	Linked Sustainable Development Goal
16. Animation Workshop	Six week course with secondary level students creating animations related to topics of sustainability	4
17. Digital Town Awards	Dingle awarded national digital town of the year	9
18. EU Living Lab Network	The Dingle Peninsula accepted into the EU living lab Network, highlighting local community approaches to sustainability and climate action	17
19. U.N. Sustainability Case Study	A UN case study created on *Dingle Peninsula 2030*	13, 17
20. School Climate Action	Working with secondary school students, 'climate hacks' were organised to devise decarbonisation pathways for transport, residential and energy production	4, 7, 13
21. KETB CommunityEducation Course	Course developed through Kerry Enterprise and Training Board in relation to community education programs	4, 13, 15
22. KETB Sustainability Strategy	Work on the Peninsula acted to inform the development of Kerry Education and Training Boards Sustainability strategy	13

(continued)

Table 7.1 (continued)

Diffusion	Sector(s)	Linked Sustainable Development Goal
23. Skyscaper.ie	Supported the emergence of a sustainable production company focused on high quality sustainable products such as clothing, office supplies and cutlery	9
24. Project 2070	Coastal adaptation project on the Maharees	13, (possibly 14 – coastal protection)

The adopted approach showcases the range of sustainability initiatives in Dingle during the project's first three years, noting early progress. However, the long-term impact and how each initiative contributes to the relevant SDGs will require extended monitoring of key indicators.

7.2 How Can We Understand the Diffusion of Sustainability in an Ongoing Community-Based Climate Action Initiative?

The discussion on the diffusion of sustainability in relation to *Dingle Peninsula 2030* highlights challenges in action research and emphasises *Dingle Peninsula 2030's* role as an intermediary. It also explores the potential for broader national and international sustainability efforts through a holistic approach aligned with the SDGs. This framework underscores the emergence of various initiatives and a collective desire for community.

Diffusion of Sustainability in Action; Challenges and Constraints

The importance of community-level actions for the delivery of climate action and sustainability has been acknowledged within policy [22, 23].

In the case of Dingle, the potential for diffusion across the community, to other communities and into national initiatives can be recognised as a means to facilitate the socio-technical transition to a low-carbon society. Investigating the diffusion of sustainability in this context can assist with this process of acknowledgement. This gives greater validity and support to community-led projects which can increase interactions between niche innovations and regime actors.

There are, however, several limitations that have been encountered which must be acknowledged. Firstly, the issue of attribution must be addressed. Aligning with the multi-level perspective [24] transitions operate across and between three levels: the niche, the regime and the landscape. Understandings of the diffusion of innovations previously outlined can be incorporated, whereby diffusion can occur from internal, external or mixed influence. Within our outline of the different initiatives influence has been attributed to *Dingle Peninsula 2030* (and within this the work of Mol Téic, MaREI, NEWKD, and ESB Networks). It is important to recognise the concept of mixed influence [25]. External factors significantly impact the emergence of new initiatives. While local projects help disseminate information, build capacity and foster trust, external influences such as landscape narratives about sustainability, policy support for initiatives and global economic drivers also play crucial roles. Therefore, any absolute claims of attribution should be approached with scepticism.

Another constraint is the relationship between the research approach taken and the established methods for investigating diffusion. When applying an engaged research methodology, whereby the research team is embedded within a 'live' project, seeking both practical problem-solving and the continued development of scientific knowledge, the ability to definitively outline the diffusion of sustainability, and within this the role of *Dingle Peninsula 2030,* is restricted (See: Chapter 9 for further discussion). For example, the impact (e.g. emissions reductions) associated key diffusions such as the community energy groups (2, 3 and 4 in Table 1) cannot be easily determined in the short-term as they have yet to bring a project to completion. Moreover, there is the issue of the embedded nature of the research, with researchers as active members of the project, contributions from the research approach and community actors become difficult to delineate.

The third significant methodological challenge is assessing the impact of various initiatives. For energy and climate goals, we can determine the CO2 emission savings associated with certain technologies. This is feasible for some technologies already implemented. For instance, we can track the number of solar PV installations, the savings from home energy improvements, and the adoption of electric vehicles [26]. However, since most of the projects listed in Table 7.1 were in early stages at the time of assessment, it was not possible to measure their impact on key indicators. Additionally, broader social benefits and capacity building—both of which are essential for the successful execution of these projects—are not easily quantifiable. By outlining the various forms of diffusion, we have established a foundation for further research. This research can build upon the analysis of the diffusion of sustainability conducted retrospectively. Alongside these diffusion patterns, we can expect the emergence of additional initiatives and projects, emphasising the significance of mixed influences as highlighted by insights from diffusion of innovation studies. What we have presented here is an active understanding of the diffusion of sustainability in real-time, which serves as a basis for ongoing study in this field, both in relation to the current project and other similar initiatives, both nationally and internationally.

Intermediaries for the Diffusion of Sustainability

The pitfalls of attribution and the need to incorporate theoretical understandings on mixed influence with regard to the diffusion of innovations are clear. Still, the importance of *Dingle Peninsula 2030* as it relates to the emergence of a range of sustainability initiatives both locally and nationally is self- evident. It was initiated as a collaborative project aiming to transition a geographic region in the Southwest of Ireland to a low-carbon, climate resilient, sustainable society by 2030. The project has sought to facilitate and support the emergence of different initiatives, and through this took an organic approach, allowing locally led initiatives to emerge rather than take a prescriptive approach.

Several different local efforts have emerged over the three-year time frame of the project such as the anaerobic digestion project, the community energy mentor course, and the West Kerry Dairy Farm SEC (see

Table 7.1). The *Dingle Peninsula 2030* partnership, while supporting these different projects, also allowed for the emergence of local actors and organisations to involve themselves in different aspects of the initiatives. The diffusion of sustainability, as a core idea across the different projects, was influenced by *Dingle Peninsula 2030*, with sustainability central to its vision. From the initial partnership a large number of groups have formed around specific projects, with many of the diffusions outlined in Table 7.1 having independent steering groups or committees. An important issue is the question of governance and coordination between the various local groups. Within policy developments, an emphasis is placed on supporting the emergence of vibrant communities that can lead the socio-technical transition within their own regional context. Projects such as *Dingle Peninsula 2030*, while acting within their own agenda and vision, can operate as intermediaries between top-down policy and bottom-up initiative with relation to the diffusion of sustainability.

Despite the embedded nature within the local regional context, the organisational structure of *Dingle Peninsula 2030* which combines local groups (Mol Téic & NEWKD) with a national body and research institute (ESB Networks & MaREI) builds the capacity to play a role between the top-down and bottom-up with relation to sustainable transitions. The diffusion of sustainability locally can be supported by the intermediary role played by regional collaborative projects such as *Dingle Peninsula 2030*. From a bottom-up perspective, intermediaries play a crucial role in stimulating and facilitating the emergence and continuity of grassroots initiatives [27]. On from this, they operate in a communicative role; "*as traditional boundaries between actor groups are being eroded or redefined, intermediaries would appear to play an important role in communicating across cultures of compliance (state), of competition (market) and of collaboration (civil society)*" ([28], p.1492). In relation to *Dingle Peninsula 2030*, the intermediate role is not the primary function, but the capacities of the collaborative grouping [2] enable it to be performed. This fits within the literature on intermediation whereby the intermediate role is played in balance with other roles and interests [29].

The overarching vision of *Dingle Peninsula 2030*, driven by the desire for a more environmentally and economically sustainable future on the

Dingle Peninsula means intermediate functions which facilitate initiatives aligned with this vision are mutually beneficial. By way of example, Mol Téic's support of the West Kerry Dairy Farm SEC in developing their application to SEAI or MaREI's support with relation to the initiation of the anaerobic digester feasibility study represents two among many when it comes to the provision of intermediary supports for emergent initiatives. The diffusion of sustainability is predicated on this intermediary role.

Diffusion Beyond the Regional Context

The national diffusion of sustainability has been referenced within the results with relation to the education initiatives and their emergent influence within the national system. Alongside this, there have been several outreach activities and events which seek to disseminate the story of *Dingle Peninsula 2030* both nationally and internationally [30]. Within an engaged research approach, the drive for societally relevant and impactful research interventions must not be restricted to the local context. MaREI has played a key role in the dissemination of findings and also in relation to network building events beyond the regional context. Mol Téic and ESB Networks have also led several events and meetings which align with these goals. Some of these are outlined as they relate to the diffusion of sustainability beyond the regional context.

The project has received extensive national exposure, including:

- Being one of six case studies featured in the National Broadcaster's Change Makers in Irish Universities television series.
- A central case study in the Irish University Association Campus Engage '*Engaged Research: Train the Trainer*' course.
- An exemplar engaged research project in the Royal Irish Academy Workshop and White Paper entitled '*Better together: knowledge co-production for a sustainable society*'.
- Knowledge sharing with SECAD Sustainable Communities Training Programme (3 × training sessions between Sept and Dec 2021).
- Presenting at Ireland's first energy planning conference organised by CODEMA in Nov 2021.

- Launch of Demographic and Socio-Economic Profile of the Dingle Peninsula by Minister for Community, Joe O'Brien.
- Recognised as a leading case study for transdisciplinary, action-orientated research in Science Foundation Ireland's annual report.

It has also made significant contributions to the international science-policy interface:

- Participation in events at the UNFCCC COP-26 Climate Conference, Glasgow 2021 including travelling via an electric vehicle to Glasgow for COP26 and documenting the journey via a blog, social media posts and a video.
- Being showcased at the EPRI European Workshop Week 2021 on Advancing Energy Communities.
- A dialogue session was held at International Sustainable Transitions (IST) conference 2020. The title of the session was '*Facilitating sustainable transitions through innovative governance*', and convened a number of panelists to share experiences from different viewpoints. These included; the manager of the Dingle Hub (Mol Téic), an Expert in Social Innovations and a representative from the UN.
- Hosting an armchair chat with members of the *Dingle Peninsula 2030* group to discuss their experiences with the project to date. The event was held in November 2020 and was attended by over 200 people, many from within the national Sustainable Energy Communities Network.
- The *Dingle Peninsula 2030* project was outlined by the UN regional information centre for Western Europe (UNRIC) as an exemplar project on their website.

For knowledge sharing, the 'learning brief' template was developed to collaboratively document reflections and learnings. These emerged as a useful means to capture important learnings on different projects or elements of the initiative in order to inform policy and community action elsewhere. They cover a variety of topics such as the 'Climate Hacks' with schools [31], the value of collaboration [4] or the trialing of electric vehicles [26].

7.3 The Importance of *Scaling Deep* to Achieve the Diffusion of Sustainability

So far, this chapter has demonstrated the wide variety of sustainability initiatives that can emerge from community-based collaborations, through outlining our research project in Dingle between the years 2018 and 2021. The *Dingle Peninsula 2030* partnership began with a very clear focus on the electrification of heating and transport, which is a central part of climate mitigation policy in Ireland. However, as illustrated through the mapping of emerging local initiatives to the SDGs, this has diffused into a rich diversity of sectors. The multi-stakeholder collaboration has acted as a springboard for local efforts, which stands to highlight the important role of intermediaries in bridging top-down and bottom-up objectives. The initial assessment of the diffusion of sustainability can provide a useful starting point for understanding the impact of such projects. Several challenges associated with this assessment of an emergent and evolving project have also been outlined. Finally, the ways in which the project has attempted to diffuse beyond the case study region demonstrates the valuable contribution of such efforts to policy and action elsewhere. Again, here the importance of intermediaries, and in particular researchers, is highlighted.

The holistic approach to sustainability as outlined through the sustainable development goals has been well represented through the *Dingle Peninsula 2030* project. The diffusion of sustainability can act as a concept which enables collaborators to track emergent initiatives across a wide scope of sectors as outlined by the SDGs. Community-led climate action projects can meet a wider spectrum of sustainability topics due to the diversity of interests and values contained within community contexts.

The *Dingle Peninsula 2030* initiative has come to an end, but the projects started here continue through the ongoing work of Mol Téic/Dingle Hub. From 2022 onwards, the Hub has been working in collaboration across several projects in relation to anaerobic digestion and circular economy approaches to biogas, energy efficiency within both the agriculture and hospitality/tourism sectors, a sustainable mobility pathfinder project, an artist-farmer partnership to imagine low-carbon farming futures and a data sharing community energy project. Across all these initiatives, the Hub is working to further develop and embed its 'scaling deep' approach to regional development. Acting both as facilitator, and in many cases initiator, Dingle Hub works as a springboard for local efforts, which stands to highlight the key role which transdisciplinary action-orientated collaborations can play in helping to support regional sustainability across a range of sectors.

Due to the clear impact of Dingle Hub's approach to regional development, there has been a great interest from a range of different stakeholders concerned with sustainability transitions. A common question is often asked: "How can we do this elsewhere?" To answer this question, we introduce the concept of "Scaling Deep" as central to the continued success and further development of Dingle Hub and its collaborative partnerships for sustainability.

The role of MaREI in the initial stages was crucial—providing capacity/resources and building capacity within Dingle Hub—through collaboration with researchers and unlocking access to resources and project funding. This partnership between MaREI and Dingle Hub continues through ongoing projects at a national and European level. The collaborative approach to working with research has enabled the hub to lead on delivering real local action that is documented and analysed while feeding into policy to have a meaningful impact.

Following on from the analysis of the diffusion of sustainability (2018–2021) here we outline the concept of "scaling deep" as an approach developed by Dingle Hub, as lead partner on the *Dingle Peninsula 2030* project, and further implemented through their ongoing work (2022–2024). This approach is situated within the literature in relation to transition studies and the literature on social enterprise and regional development. The conceptual development of the scaling deep model has been done with contributions from the chairperson (Muireann Kelliher) and manager (Deirdre De Bhailís) of Dingle Hub.

Scaling in Transitions and Social Enterprise Literature

The transition studies literature on social innovations for sustainability contains within it a number of different theoretical frameworks which can be used to systematically analyse and reflect upon the core elements of transition pathways. The environmental challenges faced in contemporary society across a broad range of areas give credence to the need for theoretical models which adequately assess transitions. Sustainability transitions do not focus on the macro- or micro- in isolation but rather are concerned with multi-dimensionality and co-evolution; multi-actor processes; stability and change; long-term process; open-mindedness and uncertainty; values, contestation and disagreement; and normative directionality shaped by public policy [32].

Transformative Innovations have been defined as multi-actor phenomena which can emerge within different institutional and social settings [33] and can include both technical and societal novelties [5]. Transformative social innovations are essential prerequisites for large systems change, such as those required in relation to sustainability. Within transition studies, the different theoretical frameworks used to analyse transition pathways provide mechanisms through which such processes can be understood. Strategic Niche Management (SNM) [34] outlines niches as spaces for experimentation that are shielded from the constraints and limitations of the established order. This space for exploring innovations is protected against the established regime's social, economic, cultural and technological parameters.

The multi-level perspective (MLP) [24] situates niches within a relational field containing three scales; the niche, the regime and the landscape offering a framework through which we can interpret the ever-changing dynamics of sustainability transitions. Within SNM, the MLP, and transitions studies are more broadly focused on understanding how isolated experimentation with new cultures norms, values and practices can facilitate systemic change. The work of Moore et al. [35] provides insights into the relationships between various scaling strategies and the system-wide effects of social change initiatives, highlighting the intricacies and complementary nature of the strategies involved in advancing large systems change. The scaling strategies of scaling up, scaling out and scaling deep are proposed.

The addition of the scaling deep classification by Moore et al. [35] (Fig. 7.2) deploys the spreading of narratives and stories to influence beliefs and behaviours, with a focus on values and mindsets. Knowledge is exchanged through participatory practices, networks, and communities of practice. Scaling deep is one among several amplification processes outlined concerning sustainability [36]. Separately to this, within the literature on social entrepreneurship, scaling deep has been introduced as a concept used to describe an approach to scaling social innovations which prioritises the improvement of current practice to positively impact upon beneficiaries. Within this the quality of impact is deemed of higher importance that the quantity of impacts. Social enterprises have a significant role to play in rural development, with the potential to function as neo-endogenous actors [37]. Such rural development actors are characterised by dynamic interactions between small regions and their larger contexts, while maintaining a strong local foundation.

The further work of Olmedo et al. [38] develops the concept of 'placed substantive hybridity', which combines a locally embedded focus within a blended approach to socio-economic relations. By using and (re)valuing (untapped) local resources and balancing them with resources from extra-local sources, as well as by accommodating and/or reacting to structural-exogenous forces based on the needs of their localities and population, rural social enterprises can thus contribute to the integrated development of their localities ([38], 0.10). A re-development of the scaling deep concept has taken place through the work of Dingle Hub. The concept builds upon the discussions outlined here about transitions management and the literature on social enterprises. It is, however, strongly informed by the empirical reality of the project. It may prove useful for other initiatives seeking to generate societal impact through collaborative partnerships for sustainability.

The Scaling Deep Model of Regional Development

Dingle Hub has enabled collaborative identification of needs and solutions within the community, infrastructure and enterprise sectors. By engaging with external actors—including government, corporates and individuals—this approach has fostered a living lab of multi-partner

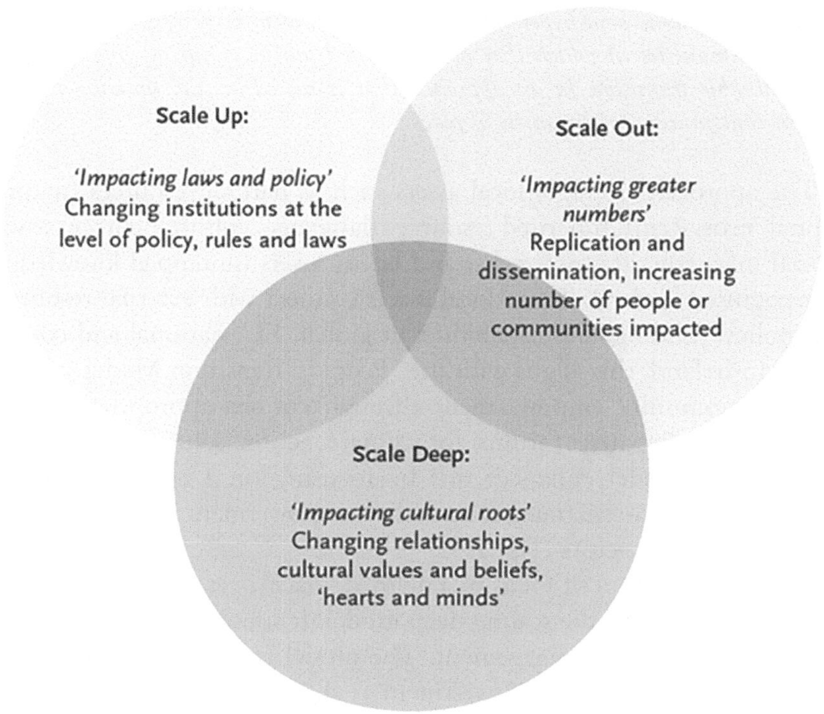

Fig. 7.2 Scaling out, scaling up and scaling deep for social innovation (from [35])

projects. The Hub's organisational framework ensures capability, inclusiveness, transparent governance and accountability. This innovative method operates at the local sub-county level, complementing traditional regional development strategies to unlock economic, social, and sustainability benefits. The strategy for regional development by the Hub is premised on the belief that current models of project-based regional development are failing to deliver to the full potential of place-based development. The Hub has pioneered their own approach to scaling deep as a mode of regional development, whereby:

> ***Scaling deep*** *is a process of regional development fusing enterprise, economic and community development where the state (in its various forms), the*

spectrum of enterprise types (from micro domestic to large corporates), and the communities in which they are embedded pursue enduring growth and sustainable transition in an approach that is rooted-in (i.e. drawing from and contributing to) the assets of place.

The approach leverages local assets such as natural resources (natural capital, ecosystem), inherited resources (language, culture), current assets (social infrastructure, networks) and latent assets (untapped knowledge, unrecognised leaders). These local assets connect with external resources like policy, funding, and capabilities at global, EU, national and county levels. In Ireland, this aligns with the "People's Transition Model", which fosters community empowerment through context-appropriate education and participative forums for climate action solutions [39]. The scaling deep model enhances this by focusing on a centralised living lab that actively orchestrates community empowerment through synergy among assets and actors (Fig. 7.3).

The concentration of focus to a defined 'place' over a sustained period enables the depth in the scaling deep model. It simultaneously allows for significant breadth of engagement. The model is asset-light with a lean and flexible resource base. Investment is directed to people, processes, networks and ecosystem capability as opposed to capital and buildings.

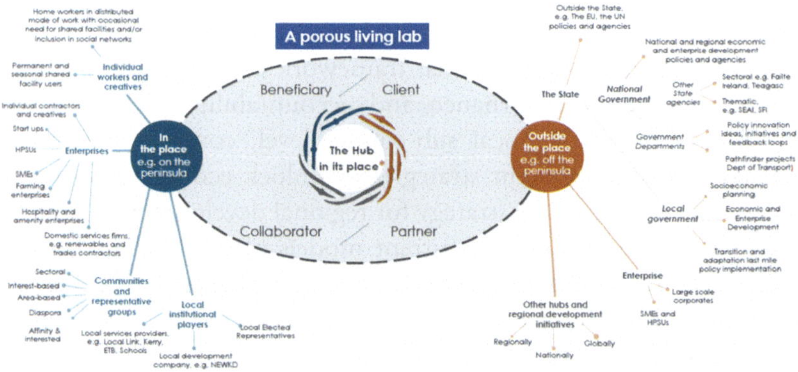

Fig. 7.3 Network of linkages for scaling deep model in the place and outside the place [40]

Despite the success of Dingle Hub to date there are still associated challenges of the scaling deep model. A lack of aligned funding mechanisms presents challenges in securing resources for core staff. The suggested core team necessary for the long-term sustainability of Dingle Hub's work would consist of a General Manager supported by a Project Manager and a Finance Manager, with further access to communications support. With an operationalised and resourced core, multiple project roles may be established to cover specific project budgets, with the ideal of having two people working on each specific project. Such attempts towards core funding remain difficult to procure, and further consideration is needed on how such core funding would relate to the registered status of Dingle Hub as a not-for-profit.

Often, project developers are wedded to traditional orthodox private-sector consultancies. Community coordination, volunteerism and engagement often go unrecognised in project funding models. There is a huge administrative burden in such an approach, and the building of a team to maximise opportunities can prove difficult. From this, the current assessment and evaluation methodologies are ill-suited for a more integrated approach to regional development, such as the scaling deep model. The concept is related yet distinct from the previously developed "diffusion of sustainability" concept [41]. While the diffusion of sustainability is applied to chart emergent initiatives emanating from the project, scaling deep is used to understand the embedding of the development into established infrastructure, be that social, cultural, technical or economic.

Despite the difficulties faced, the approach of Dingle Hub through the development of the scaling deep model combines local embeddedness, resilient leadership and sustained levels of volunteerism with access to national, regional, and local relationships of influence and the engagement of embedded researchers to support reflective learning. This model is relevant, replicable, scalable and sustainable for Dingle Hub if the associated challenged can be overcome.

7.4 Concluding Remarks

The societal impact generated through the range of projects and initiatives which have emerged on the Dingle Peninsula since 2018 is built on the establishment and continued development of the scaling deep model. This new conceptual model for regional development may act as a niche which can have broader application within the transition literature. Research on transitions in practice remains limited [32]. While the EU Climate Pact creates spaces for citizen engagement, it is coordinated through centralised points of contact at the national level. A conceptually informed approach to operationalising community empowerment has the potential to engender much more impactful climate action due to the adherence to local dynamics and characteristics it enables, while remaining connected to national and European agendas. Although established ideas concerning scaling deep exist [35], we position the experience of Dingle Hub in developing the concept in relation to regional development as an important conceptual tool for regional sustainability transitions which move beyond incumbent practices to leverage opportunities for societal impact fusing enterprise, economic and community development. Coupled with scaling deep, the diffusion of sustainability provides a mechanism to chart emergent initiatives in real-time, providing a foundation for further impact assessments to chart the value of collaborative approaches to regional sustainability adequately.

References

1. E. Boyle, Ó. Gallachóir, B.G. Mullally, Participatory network mapping of an emergent social network for a regional transition to a low-carbon and just society on the Dingle Peninsula. Local Environ. **27**, 1431–1445 (2022)

2. E. Boyle, C. Watson, C. McGookin, A. Deane, D. Bhailís, C. McElligott, S. O'Hara, B. Tuohy, B.Ó. Gallachóir, G. Mullally, Reflecting on a collaborative approach to a regional sustainability transition: Dingle Peninsula 2030. Reflect. Pract. **22**, 416–430 (2021)
3. C. McGookin, T. Mac Uidhir, B.Ó. Gallachóir, E. Byrne, Doing things differently: Bridging community concerns and energy system modelling with a transdisciplinary approach in rural Ireland. Energy Res. Soc. Sci. **89**, 102658 (2022)
4. C. Watson, E. Boyle, C. McGookin, A. Deane, D. Bhailís, C. McElligott, S. Oara, B. Tuohy, B.Ó. Gallachóir, *The value of collaboration. Dingle Peninsula 2030 learning brief series* (MaREI Centre, Cork, 2021)
5. D. Loorbach, J. Wittmayer, F. Avelino, T. Wirth, N. Frantzeskaki, Transformative innovation and translocal diffusion. Environ. Innov. Soc. Transit. **35**, 251–260 (2020)
6. G. Rossman, J.C. Fisher, Network hubs cease to be influential in the presence of low levels of advertising. Proc. Natl. Acad. Sci. **118**, 2013391118 (2021)
7. T.W. Valente, E.M. Rogers, The origins and development of the diffusion of innovations paradigm as an example of scientific growth. Sci. Commun. **16**, 242–273 (1995)
8. T. Kluvankova, M. Nijnik, M. Spacek, S. Sarkki, M. Perlik, R. Lukesch, M. Melnykovych, D. Valero, S. Brnkalakova, Social innovation for sustainability transformation and its diverging development paths in marginalised rural areas. Sociol. Rural. **61**, 344–371 (2021)
9. H.A. van den Heiligenberg, G.J. Heimeriks, M.P. Hekkert, R.P. Raven, Pathways and harbours for the translocal diffusion of sustainability innovations in Europe. Environ. Innov. Soc. Transit. **42**, 374–394 (2022)
10. E.M. Rogers, Diffusion of innovations: modifications of a model for telecommunications. In: *Diffus. Von Innov. Telekommunikation* (Springer, Berlin, 1995) pp. 25–38
11. A. Palloni, Diffusion in sociological analysis. In: *Diffus. Process. Fertil. Transit. Sel. Perspect.* (2001), pp. 67–114
12. G.W. Clough, J.L. Chameau, C. Carmichael, Sustainability and the University. Presidency **9**, 30 (2006)
13. R. Hölsgens, S. Lübke, M. Hasselkuß, Social innovations in the German energy transition: an attempt to use the heuristics of the multi-level perspective of transitions to analyze the diffusion process of social innovations. Energy Sustain. Soc. **8**, 1–13 (2018)

14. G.M. Eichler, E.J. Schwarz, What sustainable development goals do social innovations address? A systematic review and content analysis of social innovation literature. Sustainability **11**, 522 (2019)
15. R. Bali Swain, F. Yang-Wallentin, Achieving sustainable development goals: predicaments and strategies. Int. J. Sustain. Dev. World Ecol. **27**, 96–106 (2020)
16. S. Nicolai, C. Hoy, T. Berliner, T. Aedy, *Projecting Progress: Reaching the SDGs by 2030* (Overseas Development Institute, London, 2015)
17. J.D. Sachs, From millennium development goals to sustainable development goals. The Lancet **379**, 2206–2211 (2012)
18. S.F. Everton, S. Pfaff, Historical and comparative research on social diffusion: mechanisms, methods, and data. Soc. Sci. Hist. **46**, 431–472 (2022)
19. J.S. Horng, C.H. Liu, S.F. Chou, C.Y. Tsai, Y.C. Chung, From innovation to sustainability: sustainability innovations of eco-friendly hotels in Taiwan. Int. J. Hosp. Manag. **63**, 44–52 (2017)
20. R. Chaudhary, C. Kumar, Determinants of diffusion of environmental sustainability innovations in hospitals of Bihar state in India. J. Glob. Respons. **12**, 76–99 (2020)
21. K.R. Smerecnik, P.A. Andersen, The diffusion of environmental sustainability innovations in North American hotels and ski resorts. J. Sustain. Tour. **19**, 171–196 (2011)
22. Department of Communications, Climate Action & Environment, Climate Action Plan - To Tackle Climate Breakdown., Department of Communications, Climate Action and Environment, Dublin (2019). https://www.dccae.gov.ie/en-ie/climate-action/publications/Pages/Climate-Action-Plan.aspx. Accessed 1 July 2020
23. A. Caramizaru, A. Uihlein, *Energy communities: an overview of energy and social innovation* (Publications Office of the European Union, Luxembourg, 2020)
24. F.W. Geels, Technological transitions as evolutionary reconfiguration processes: a multi-level perspective and a case-study. Res. Policy **31**, 1257–1274 (2002)
25. T.W. Valente, Diffusion of innovations and policy decision-making. J. Commun. **43**, 30–45 (1993)
26. E. Boyle, C. McGookin, C. Watson, A. Deane, B. Gallachóir, *ESB Networks Dingle Project: Activating Energy Citizenship Through Collaboration, Participation, and Technological Trials* (MaREI Centre, Cork, 2021)

27. C. Bird, J. Barnes, Scaling up community activism: the role of intermediaries in collective approaches to community energy. People Place Policy Online 8 (2014)
28. T. Moss, Intermediaries and the governance of sociotechnical networks in transition. Environ Plan A **41**, 1480–1495 (2009)
29. W. Kanda, M. Kuisma, P. Kivimaa, O. Hjelm, Conceptualising the systemic activities of intermediaries in sustainability transitions. Environ. Innov. Soc. Transit. **36**, 449–465 (2020)
30. MaREI Centre, Dingle Peninsula 2030 (2022). https://www.marei.ie/dingle-peninsula-
31. C. McGookin, S. Moriarty, D. Bhailís, B.Ó. Gallachóir, *Climate Hacks. Dingle Peninsula 2030 Learning Brief Series* (MaREI Centre, Cork, 2020)
32. J. Köhler, F.W. Geels, F. Kern, J. Markard, E. Onsongo, A. Wieczorek, F. Alkemade, F. Avelino, A. Bergek, F. Boons, L. Fünfschilling, An agenda for sustainability transitions research: state of the art and future directions. Environ. Innov. Soc. Transit. **31**, 1–32 (2019)
33. F. Avelino, J.M. Wittmayer, Shifting power relations in sustainability transitions: a multi-actor perspective. J. Environ. Policy Plan. **18**, 628–649 (2016)
34. R. Kemp, J. Schot, R. Hoogma, Regime shifts to sustainability through processes of niche formation: the approach of strategic niche management. Technol. Anal. Strateg. Manag. **10**, 175–198 (1998)
35. M.L. Moore, D. Riddell, D. Vocisano, Scaling out, scaling up, scaling deep: strategies of non-profits in advancing systemic social innovation. J. Corp. Citizsh. **58**, 67–84 (2015)
36. D.P. Lam, B. Martín-López, A. Wiek, E.M. Bennett, N. Frantzeskaki, A.I. Horcea-Milcu, D.J. Lang, Scaling the impact of sustainability initiatives: a typology of amplification processes. Urban Transform. **2**, 1–24 (2020)
37. L. Olmedo, M. O'Shaughnessy, Community-based social enterprises as actors for neo-endogenous rural development: a multi-stakeholder approach. Rural. Sociol. **87**, 1191–1218 (2022)
38. L. Olmedo, M. Twuijver, M. O'Shaughnessy, A. Sloane, Irish rural social enterprises and the national policy framework. Administration **69**, 9–37 (2021)
39. S. McCabe, The people's transition: community led development for climate justice. Found. Eur. Progress. Stud. Bruss. (2020)
40. E. Boyle, D. De Bhailís, M. Kelliher, Scaling Deep : Societal impact through a new approach to regional development. Soc. Impacts **4**, 100083 (2024)

41. E. Boyle, C. McGookin, D. Bhailís, B.Ó. Gallachóir, G. Mullally, The diffusion of sustainability and Dingle Peninsula 2030. UCL Open Environ. **4**, e052 (2022)

8

Community Engagements Approaches

Connor McGookin and Evan Boyle

8.1 Understanding the Different Ways People Can Be Engaged in Research

There is a wide range of engagement activities that can be undertaken by academics, which may be classified based on the level of control over the research outcomes and design. The 'Public Engagement Onion' is a useful framework developed by the Wellcome Trust, it puts engagements into three categories [1]:

- Informing—one-way flow of communication, usually for the purpose of informed decision-making, awareness-raising or educating, e.g. presentations at public events, policy submissions, online material.
- Consulting—two-way flow of communication, surveys, interviews, or workshops used to elicit opinions; participants have an opportunity to shape the research results but not the research questions or objectives.
- Collaborating—open and transparent communication throughout the research process, participants are given the opportunity to shape research questions and direction throughout the duration of the project.

C. McGookin and E. Boyle, *Engaging Communities in Climate Action*, Energy, Climate and the Environment, https://doi.org/10.1007/978-3-031-93498-8_8

During the *Corca Dhuibhne/Dingle Peninsula 2030* project, our research team employed a variety of activities across each of these different engagement categories. All three forms of engagement are important in different contexts.

- *Informing* is useful for sharing learnings, insights, and recommendations arising from the project and also as an initial entry point into the community.
- *Consultations* are a means of building an understanding of the needs and priorities of people from the area.
- *Collaborating* with the project partners and co-producing outputs enables the research to be embedded within the community and to actively support activities 'on the ground'.

In mapping the various ways that we contributed to the project (Chapter 9, Sect. 9.1), these three categories were extended into six. As outlined in Table 8.1, this was done to make distinction between activities within the study region versus further afield, individual versus group consultations, and ways in which we were collaborating.

8.2 Informing

This section covers both public events held in the case study region and efforts to share the learnings with national and international networks.

8.2.1 Climate Hack

Background

The event was run as part of Science Week 2019 in collaboration with the Dingle Creativity and Innovation Hub. It involved a two-hour workshop with fourth- and fifth-year students in the three secondary schools on the Dingle Peninsula; Pobalscoil Chorca Dhuibhne, Coláiste Íde, and Meanscoil Nua an Leith-Triúigh. The format of the event is based on

Table 8.1 Mapping our activities to the engagement onion, adopted from [1, 2]

Level of participation	Sub-levels	Explanation
Informing	Dissemination of findings	Sharing learning from the project with interested groups: policymakers, academics, other communities, etc.
	Stimulating thinking on the Dingle Peninsula	Events like family days organised on the Dingle Peninsula to promote the project
Consulting	Understanding individual perspectives	Interviews or surveys conducted
	Facilitating dialogue	Group workshops and discussions
Collaborating	Collaborating	Co-hosting events, working on joint funding applications, and sitting on steering groups or attending committee meetings
	Co-producing	Co-authored outputs; journal articles, reports, and learning briefs

the 'serious game' approach to scientific communication. A serious game is an interactive approach that is designed with the intention to teach rather than purely entertain. It is a very commonly used method of engagement in climate change research [3–5]. These games can help raise awareness, build capacity for problem-solving, and provide a space to explore potential futures [6]. A well-known example of the serious game approach is the World Climate Simulation, which asks participants to take on the role of UN negotiators and agree global climate change policy (Chapter 6, Sect. 6.2).

The idea of the 'climate hack' template developed for the Dingle Peninsula was inspired by the work of Krzywoszynska et al. [7],Thomas et al. [8], and Volken et al. [9]. These studies all used some form of information sheets/cards to help participants make informed decisions about the future energy system. A similar approach was applied with the 'climate hack', where an introductory brief on the current energy balance and associated CO_2 emissions, and an information sheets on the options available to decarbonise heating, electricity and transport were prepared.

The purpose of providing this information was firstly to highlight the challenges facing the Dingle Peninsula and secondly to ensure that the students could make well-informed decisions about the future of their energy system.

Aims and Objectives

The overall aim of the event was to explore options for a sustainable energy future with young people in the area. By giving them usable information and asking them to assume the role of a town planner, an interesting insight was gained into what parts of the energy transition they viewed as most important and their thoughts on the different solutions available.

The key objectives may be summarised as follows:

- Inform students of the current energy demand and supply in their area, highlighting large sources of CO_2 emissions such as private car travel or home heating.
- Give students the necessary data on electricity, heating, and transport energy options so they can make informed decisions about the future.
- Develop local energy projects guided by young people's vision for a more sustainable future.

It was decided that the event would not be run as a competition but rather that following the event the most enthusiastic group from each school would be asked to develop their idea further and prepare a poster presentation to display at the public launch of the Dingle Peninsula Energy Master Plan. In the end, the best idea from each challenge was chosen, giving a presentation on a community solar farm, public transport, and alternative sources of heat (anaerobic digestion and hydrogen from electrolysis).

How It Worked

Preparation—One week prior to the event, the introductory brief was sent out to schools. This gave an overview of energy in the region, how much this was costing the average household, and the associated CO_2 emissions. It then prompted the students to investigate three key challenges facing the area; private car travel, home heating, and electricity supply. In groups, the students were asked to choose one of these challenges or propose their own. A number of useful links were provided so that students could conduct their own piece of research in preparation for the event. The resources needed are summarised in Table 8.2.

The information on energy and alternative options for heating, transport, and electricity was primary based on my work to develop a baseline energy balance and energy scenarios for the region (Chapter 6).

The Event ran for 2 hours as follows;

- 5 mins—Initial welcome address from Dingle Creativity and Innovation Hub
- 10 mins—Introductory presentation from MaREI on the Dingle Peninsula energy balance and low-carbon solutions in heating, transport, and electricity.
- 1 hr 15 mins—Students were given the information sheet for their chosen challenge as well as an answer template with four sections to fill in; CO_2 savings, cost estimate, potential barriers, and benefits of their proposal. The data needed and example calculations were provided on

Table 8.2 What was needed to organise the Climate Hacks

Person-hours	
Preparation*	24
Event	18
Total	42
Cost	
Translation**	€150
Printing and Co-ordination	€250
Video	€1100
Travel and Accommodation	€160
Total	€1660

* Preparation of the material and event organisation only covers researcher input,excludes preparation time of Dingle Hub, teachers and students

** Corca Dhuibhne is aGaeltachtregion, so all material was bilingual

the information sheet. In addition, tablet computers were available for further research.
- 30 mins—Each group was asked to present their idea(s).

Lessons learnt

- Preparation is important—if students are given time to form groups and decide what they want to work on before the event then they get a lot more out of it. The initial information sheet should be sent to schools at least one week before the event.
- Demonstrator ratio—In order to ensure the students are given adequate attention and maintain focus throughout the exercise, it is best to have at least one demonstrator for every three groups of students.
- Not everyone is an engineer—The calculations should be simple, with easy-to-follow instructions. It is only necessary to give an overview of the relative cost of solutions in order to prompt a discussion about trade-offs like the fact that micro (domestic-scale) wind energy might be more popular but is not as cost effective as large (commercial-scale) projects.
- Focus on creativity—The solutions sheets should focus on developing an idea rather than calculating CO_2 savings or a cost estimate. The most important element of the event is to give young people an opportunity to make suggestions, not give them a math lesson.

8.2.2 LED Bulb Swap

Background

The idea for the 'LED bulb swap' came from a LECo (Local Energy Communities) project tour to the Rhein-Hünsruck District in Germany, hosted by the renewable energies' agency in November 2018 [10]. Since early 2012, the district has been a renewable electricity exporter, producing more than its total electricity demand, an important milestone on the path to a 100% renewable energy system [11].

The purpose of the trip was to see how the region is transforming the €290 million previously spent on energy imports into local jobs and regional value [10]. The area of Neuerkirch had some particularly good examples of how profits from the export of electricity could benefit the community. They purchased a community bus that operates a taxi service and looks after the senior citizens. In addition, when the heat network was being developed, 300mb fibre internet was installed into every house. It was also here that the idea of an LED bulb swap emerged. Out of the 105 houses in the area, 87 had all their lightbulbs replaced with LED bulbs, 1,452 lightbulbs were replaced completely free of charge. The cost to replace them was €6,300 but the cost saving per year was estimated at €8,300. Across 40 communities within the Rhein-Hünsruck District there have been 14,000 bulbs from 1,100 households exchanged free of charge.

Aims and Objectives

The overall aim of the event was to begin a discussion on energy demand in people's homes. By demonstrating how something as simple as changing to LED bulbs can have a significant impact on the electricity needed to provide lighting.

The key objectives may be summarised as follows:

- Inform people of the options for improving the energy efficiency and comfort of their homes.
- Launch the Credit Union's 'green loan', with special low rates for the financing of home energy improvements.
- Highlight the ongoing activity in the area being organised through the Dingle Peninsula Sustainable Energy Community (SEC).
- Prompt people to fill in a survey on household energy performance as part of the development of an energy master plan for the area.

How It Worked

The event was run on a Thursday evening for two hours in April 2019. People could bring old incandescent bulbs and exchange them for new LEDs, free of charge. There was a limit set of four bulbs per person. Alongside the bulb exchange there was a demonstration of how much less energy a LED needs versus an incandescent bulb, several other 'green' educational electricity sets were available for children to play with, and there was information stands from the Dingle Peninsula Sustainable Energy Community, Credit Union, MaREI, ESB Networks, and Local Link.

Lessons Learnt

- People are generally interested in ways to improve the energy efficiency of their homes but are unable to make the significant investment that may be required for a retrofit.
- Careful bulb selection—out of the 4 LED bulbs available for swap only one was a bayonet fitting, but this was by far the most popular choice. In addition, the lumens (brightness) of the bulbs being given out were much higher than they needed to be.

8.2.3 Learning Briefs

Background

The *Dingle Peninsula 2030* 'Learning Briefs' are short documents to share learnings from particular activities, projects, or events. It provides a simple, clear summary of important topics.

The learning briefs:

- provide context and background
- document what happened and how it happened
- explain the roles of the different participants

- highlight what worked, what didn't work, and how improvements could be made
- make recommendations for policy or other key actors

Aims and Objectives

Learning briefs are produced relatively quickly compared to other academic outputs like journal papers, and thus are a way of collecting and presenting data in close to real-time. The overall aim is to capture the learnings in a format that is accessible and useful to the participants involved and to other communities, groups, researchers, companies, agencies and policymakers. The process of writing a learning brief involves critical reflection and deliberation. They thus also provide a useful form of evaluation, which helps us as researchers to learn and improve.

How It Worked

The writing process was coordinated by our research team, but documents are co-created with other members of the partnership. Generally, we would prepare a first draft to share with the relevant co-authors who had been involved with the event or project and then host a meeting to discuss feedback on it.

Lessons learnt

- The learning briefs were a great way to share recommendations and lessons emerging form the project. Although quantitative metrics cannot tell us exactly how useful the briefs were, they nonetheless indicate that the template was an impactful means to reach a wide audience. The Energy Master Plan learning brief was viewed over 4,000 times between 2020 and 2021, while the journal paper on the same topic was read just 94 times.
- Writing needs to be lead by the researchers—our project partners had very limited time to input on the drafts, and preferred giving comments on them rather than drafting text.

- Hosting a meeting is a good way to gather feedback, it sets a timeslot for people to review the material and provide input without having to make their own edits on draft text.
- The learning briefs offered a means to disseminate elements of the project that would not fit into an academic paper and also a means to share insights from accompanying papers in a more digestible format.

8.3 Consulting

This section covers the various community consultations that took place; it covers a range of different workshops held as well as interviews conducted.

8.3.1 Student Workshop

Background

Following advice from some colleagues, students in University College Cork that are from the Dingle Peninsula were identified as a good opportunity to start a dialogue about the project and get a sense of the community concerns. Working with a group that is based on campus removes the need to travel, and there are plenty of venues readily available for hosting a workshop.

Aims and Objectives

The goal was to meet with some of the students to understand what the priorities and concerns of young people from the area. What are their hopes for the future? Do they plan to return to the area after college? The discussion was also an opportunity to present our plans for the project, get feedback on it, and advice on how to engage with the wider Dingle Peninsula community.

How It Worked

Using some suggested contacts from our local partners, the process began by reaching out to students and inviting them to go for a coffee. These informal discussions were held with four students. It was by no means an interview, but rather just a conversation about what they were studying, their interests, plans for the future, and what they liked about their home. Two came from Dingle Hub's Youth Advisory Board, and the other two were suggested by these initial contacts.

A larger workshop was then organised on campus. It was split into three parts: 1) a survey of the students, answered individually first and then discussed as a group, 2) a SWOT analysis of the peninsula, and 3) a presentation on the *Dingle Peninsula 2030* project and our plans for research. We got a small budget to cover pizza during the break, otherwise the room was free, and I facilitated everything. There were 13 attendees, mainly in the later (3rd and 4th) years of their degrees.

Questions Posed

- What do you like about being from the Dingle Peninsula?
- Anything you would like to change about the Dingle Peninsula in the short/long term?
- Please write a paragraph describing the Dingle Peninsula you would like to see in the future. Think about questions like: How has the community developed? What jobs are people doing? How do people get around? What are the houses like? How has the landscape changed?
- Do you have any further comments/suggestions on the initiatives developing on the Peninsula? Is there anything you think Dingle Hub or ESB Networks should be doing or could be doing better? Or any other stakeholders you think should be involved?
- Do you have any further comments/suggestions on the proposed research? Is there anything you think we should be doing or could be doing better?

Summary of findings

Theme 1: Dingle is a small, tightly knit community	Key questions arising
Small town culture • Everybody knows each other • It's a very safe place to live • It's a very diverse community, plenty of well-integrated blow-ins • There is a good sense of community • People of Dingle are hardworking, proud of the area and committed to promoting / preserving it • It thrives because of; *"The hardworking nature and innate want of people who live in Dingle to make Dingle a better place"* **Vibrant area** • Great tourist destination • Plenty of summer job opportunities *"For a town of its size it has many opportunities other towns don't have due to the tourism industry"* • Many festivals; film, food, arts music – always something happening *"As a young adult, there is loads of reasons to come home every weekend, e.g. food festival, Other Voices,* etc.*"*	How can Dingle protect this strong sense of community into the future? • Handling the integration of more people *"Increasing number of blow-ins may change the character of the community"* • Planning infrastructure / development appropriately
Theme 2: The Dingle Peninsula is a beautiful landscape	**Key questions arising**
Place of natural beauty • Great scenery • Very natural place *"Dingle is still more in touch with a traditional way of living"* **Protecting the landscape** • Increasing tourism shouldn't threaten landscape *"I'd like the landscape to stay the same, as natural as possible"* *"Landscape untouched. Beaches, wildlife being protected"* • Increase in littering if tourism keeps expanding	How can Dingle ensure that increasing number of tourist and demand for infrastructure/ development doesn't impact on the landscape?
Theme 3: Lack of Opportunities for Youth	**Key questions arising**

(continued)

(continued)

Theme 1: Dingle is a small, tightly knit community	Key questions arising
The "brain drain" • Limited opportunity outside of hospitality and agriculture • Need for higher scale job opportunities *"steer our reliability away from tourism"* *"I would like to see Dingle as a place that people who grew up in Dingle could come back to and use their qualification"* • Continued unsustainable growth in the tourism sector makes the area increasing less attractive to native people seeking to work outside of hospitality **Lack of access to Wi-Fi** • There needs to be a reliable internet connection • More modern facilities like the Hub available *"love to see more work spaces like Dingle Hub from which people can pursue international careers"* **Lack of long-term rentals** • Limited amount of long-term rentals available • Young people choosing to stay in Dingle can't find anywhere to live *"There are far too many short-term houses being let, we need long term letting"*	How can the Dingle Creativity and Innovation Hub (expand its services to offer young people the chance to work remotely? • Promoting the notion of working from home • Drawing in interested companies • Expanding on Wi-Fi access
Theme 4: Planning / Infrastructure	**Key questions arising**

(continued)

(continued)

Theme 1: Dingle is a small, tightly knit community	Key questions arising
Remoteness • Small isolated towns scattered around the peninsula at risk • Lack of facilities for young people *"not a lot of things for young people (secondary school students) to do outside of school"* **Essentially no public transport, heavy reliance on car** • Need for better planned infrastructure • Traffic during tourism season a problem *"like to see Dingle as a place that is free of traffic congestion"* *"No more widening of roads"* • With more transport options *"The infrastructure is as good as it should be, the only addition I could see is bicycle lanes",* *"a smart system of transport for the elderly"* **Room for improvement** • Small farmers, eager to explore new sources of income • Very dispersed housing/buildings, leaving plenty of room for the development of new industry • However, some also highlighted a fear of developers; *"who solely see Dingle as an opportunity to make money"* • Moreover, noted the need to protect less developed areas in the interest of protecting the nature beauty *"Not allowing planning permission outside of towns, ruining the landscape"*	Could the future developments be better planned, considering the concerns of locals?

Lessons Learnt

- Be very clear what the purpose of the workshop is. Some of the students that attended were surprised when they were asked to fill in the ethics form and survey. They had expected to hear a short presentation on the project, but hadn't realised we would also be asking them a series of questions.

- Informal settings like going for a coffee are a good way to build an understanding of people's perspectives. However, it provides limited material for the research as the conversations were not properly structured or recorded.
- Asking the students to first fill in the questions individually before discussing as a group ensured that all voices were heard despite some people not speaking up during the event. After the workshop when typing up the survey answers, there was some issues written down that had not been raised during the discussions.

8.3.2 Community Meetings

Background

The idea of undertaking a Demographic and Socio-Economic study of the Dingle Peninsula arose from Social Inclusion and Community Activation (SICAP) focused work and ongoing discussions within NEWKD's Corca Dhuibhne's Local Area Advisory Committee (LAAC) in 2019. The Corca Dhuibhne LAAC is a voluntary committee and comprises 14 geographic and 'issue based' representatives who live on the peninsula, and who are actively involved in contributing to the development of their community. A LEADER grant and co-funding from MaREI was secured to contract Breandán Ó Caoimh to undertake the work. While Breandán presented the demographic and socio-economic trends of each local area, I presented the energy facts.

Aims and Objectives

The purpose of these meetings was to give people key information about their area and facilitate the development of an evidence-based plan for a sustainable future; across social, economic, and environmental concerns.

The meetings were timed to align with the drafting of the County Development Plan, which is the forward planning conducted by Irish local authorities every six years.

In summary, the goal was to:

- Provide communities with information on the key social, economic and environmental challenges facing them
- Listen to what each community and organisation across the peninsula thinks and knows about the social, economic, and environmental challenges facing them
- Support the emergence of working groups across the peninsula around specific projects or topics

How It Worked

The Dingle Peninsula is made up of 8 parishes, each with a parish council. The council network is a group of volunteers, supported by NEWKD to deliver some community services. There were two elements to this study: desk-based research on relevant baseline indicators for each of the 8 communities, community consultations through a survey and 15 public meetings.

The desk-based research drew together information on the demographic and socio-economic profile of the area [12], and the analysis undertaken to inform the Dingle Peninsula Energy Masterplan (Chapter 6). Together, these research strands provided a quantitative description of the peninsula in respect of such variables as population numbers, household and family types, age of houses, heating fuels used and cost of fuel.

In addition to desk research, community consultations took place through 15 public meetings in late 2019 and early 2020, in the 8 geographic communities throughout the peninsula. Approximately 400 people attended these meetings. Members of the community were given a brief presentation on the findings of the desk research, and they were then invited to give feedback on the implications for their community, discuss the strengths and challenges of their community, and what could be done. The first round of meetings focused on the current challenges, while the second round looked to build recommendations for how to address them.

Lessons Learnt

Meet people where they are at:

- NEWKD worked with local community representatives to host events in locations and at times which were convenient for residents to attend.
- Tailor the language and material to the audience. The understanding of a sustainable community must be broadened beyond climate imperatives. For rural communities, a sustainable future means long-term social, economic, and environmental resilience.
- Allow time for less formal discussions to take place during tea/coffee breaks.

Tap into local networks:

- Use local contacts. NEWKD was able to use its contacts in local media to publicise these community meetings, and thereafter to give updates on outcomes at meetings.
- NEWKD was encouraged to undertake this study by its LAAC members and the broader community. Having 5 previous years of working with and supporting local groups provided credibility to engage successfully with communities throughout the peninsula.

Be open and transparent:

- All of the material and outputs should be made available online.
- Have multiple options for participating, such as allowing for written submissions if people can't attend the meetings, and surveys, as well as in-person meetings.

8.3.3 Re-Imagine Dingle Peninsula

Background

Following the onset of the pandemic, Dingle Hub wanted to host an online workshop to explore how the region could 'build forward better' after the COVID-19 pandemic. With reduced international visitors and periods of complete lockdown, the hospitality sector that the area heavily relies on suffered a large loss of business. However, there were also positive signs that people had made their temporary holiday homes a permanent home during the pandemic, and the Hub was keen to identify ways that this trend could be continued.

Aims and Objectives

The purpose of the 'Re-imagine Dingle Peninsula' workshop was to identify projects of interest from key community members in the hope that they would form working groups and continue to develop the ideas.

How It Worked

The groups were selected by Dingle Hub to try and match people of similar interests. The participants were invited from different business sectors, and also across young and old age groups. In total 28 participants were invited and put into 6 groups.

The workshop was facilitation by Imagination of Things, and a variety of different tools were used to host the event. It took place over a weekend, with two sessions held: 1) on Friday afternoon to generate ideas and 2) a feedback workshop on Monday to share the ideas developed with the other groups. Over the weekend, the participants were given funds to cover a meeting at a coffee shop to discuss their idea further and prepare a presentation for the 2nd session.

- Google Meet—video conferencing for overall event and break out groups
- Miro board used to gather the ideas

- Google Slides/Docs used to share instructions for the exercises and prepare presentations on the project ideas

Lessons Learnt

- Working online—Although some people had difficulty editing the Miro board and moving around, the vast majority worked away very effectively. It was a great way to capture everyone's input at the same time. However, it is important to have note takers or facilitators on hand to document the input of people that find navigating these online environments difficult.
- You need a lot of facilitators—ideally each breakout room needs a facilitator. Workshops often run on very tight schedules since you don't want to keep people for too long. Especially in an online environment it is best to limit sessions to 90 minutes or 2 hours. Therefore, you need to squeeze a lot out in short break out sessions of 20–30 minutes. Facilitators are important to keep the conversation focused and ensure the instructions are properly understood.
- Computer and internet access—some older participants did not have access to a computer or laptop, also several people live in areas that have no Wi-Fi coverage. Dingle Hub had spare laptops available for those that needed it, and also opened space for people to get internet access. These are important considerations to ensure people aren't excluded.

8.3.4 Participatory Mapping

Background

At the early stage of the project, an initial mapping exercise was completed with key community representatives related to the project. Through embedded ourselves in the community prior to the commencement of the project, we were able to identify key community organisations and actors. From this, we arranged a workshop with four key community partners. This acted as a first interaction with some of the community organisations and it was an opportunity for them to strategically think about their own individual goals and the capacities and supports required to achieve these objectives.

Aims and Objectives

We sought to bring together key community actors to understand the local and national social network surround their goals. We aimed to provide strategic insight on how to build capacity and leverage their networks to achieve their goals. This offered the potential to highlight opportunities for the participating organisations to work together on shared goals.

How It Worked

This low-tech intervention used simple supplies which helped us to easily organise the event. A white board was placed face down on the table and different coloured markers and sticky notes were provided to facilitate the mapping. Two facilitators conducted the workshop.

- In step 1, participants were asked to collaboratively list all the different stakeholders currently involved in the range of initiatives being discussed. Colour coding was used to define those within the region (local) and those outside the region (external).
- In step 2, participants were then asked to outline their individual goals from the perspective of the community organisation they represented. Colour coding was used for short (-1 year), medium (1–5 years), and long (5 + years) term goals.
- In step 3, discussion was facilitated to outline the key supports provided (or which could be provided) in order to achieve these different goals. Five different supports were outlined and coloured coding was applied using the coloured markers.
- In step 4, the stakeholders listed were connected to individual goals using the supports outlined.

- Finally, in step 5, a discussion was facilitated to reflect upon the network which emerged (Fig. 8.1).

Lessons Learnt

- This is beneficial as a process for inclusive participation to achieving just transitions. However, the data represented through this process is only individual interpretations of networks in a given context, at a certain time, despite the collaborative mechanism deployed.
- Often such methods include the use of perceived influence dynamics related to different organisations, i.e. who has more or less power. Due to being an early intervention with new partners we remained cognizant of the potential divisiveness of this approach. When undertaking engaged research, you must balance the optimum research approach with the collaborative partnership.

Fig. 8.1 Picture of Participatory network mapping exercise adapted from Net-Map Toolkit

8.3.5 Ambassador Programme in ESBN Dingle Project

Background

To better understand the challenges faced by individuals and households in transitioning away from solid fuels, the ESB Networks Dingle Project initiated their Ambassador Programme on the peninsula, whereby local people applied to integrate new technologies into their homes and businesses, to investigate the lived experience and to share learnings from their participation in the project. Five ambassadors were selected and as part of the trial three retrofits were completed (two domestic and one commercial). The deep retrofits included internal and external insulation, the installation of new doors and windows, air source heat pumps and ventilation systems. All five Ambassadors had seven solar panels installed alongside a residential scale battery and an EV for a 12-month period. We followed the five ambassadors over the three-year trial to understand their experience of integrating new energy technologies into their lives.

Aims and Objectives

This research sought to understand the experience of the five ambassadors when integrating new technologies into their lives. The future energy system requires more active engagement of 'customers'. Coupled with this, new technologies can present unexpected challenges for users to overcome. Through following the process over a three-year period, we were able to analyse these developments.

How It Worked

Twenty interviews were conducted over four rounds from 2018 to 2021. After each round, following transcribing the recordings of interviews, the data was analysed thematically and coded. Once coded, categories were created around different themes where similarities and overlaps were present across different interviews. Questions for each round of interviews was informed by the previous round, with supporting insight from ESB Networks and media outreach undertaken by the ambassadors.

Lessons Learnt

- The ESB Networks Ambassadors, throughout the three-year process, have emerged as examples of energy citizens, becoming more versed in energy-related topics and engaging publicly through media outreach and public presentations to share their experiences of participation in the trial. These Ambassadors now see themselves as having a voice within the community in relation to energy.
- The lack of available quantitative data for the Ambassadors to assume a role as energy citizens throughout the trial and to be a trusted voice in the community. The COVID-19 pandemic delayed the implementation of technologies with knock-on effects around the rollout of an app for participants to have direct interaction with the quantitative information.
- Working in partnership with both citizens and industry partners presents challenges in managing timelines and expectations. Unexpected delays in relation to procurement (impacted by COVID-19) meant our qualitative data could not be analysed in relation to long-term quantitative data on energy usage. This severely limited the rigour or the research.

8.4 Collaborating

A central element to the approach adopted by MaREI was that external stakeholders would not only provide input to shape the results but also be part of the research design and decision-making process. The forming of partnerships was seen as a matter of best practice. It was critical to ensuring the research had impact but was not without its challenges (Chapter 9).

8.4.1 Dingle Peninsula Sustainable Energy Community

Background

In April 2018, the Dingle Peninsula Sustainable Energy Community (SEC) was established by Mol Teic (Dingle Creativity and Innovation Hub). The SEC network is coordinated by SEAI, with the aim of encouraging and supporting the involvement of communities in the transition to a low-carbon future [13]. A vital first step in the development of an SEC is to understand the community's current energy balance within the SEC area and to develop a 'Register of Opportunities' that identifies what specific projects could be undertaken to reduce the energy demand and switch to renewable energy.

In October 2018, following a tendering process, the Dingle Peninsula SEC commissioned the production of an Energy Master Plan for the Peninsula, at a cost of €20,000. The work was completed in June 2019, and, following review by the Steering Committee and additional analysis from the MaREI Centre [14], an Energy Master Plan was subsequently published in February 2020 [15].

Aims and Objectives

The Energy Master Plan (EMP) provides an overview of the community's current energy demand, as well as the options available for reducing energy usage and switching to renewable energy sources. The purpose of the EMP is to begin a conversation about the local energy system and the necessary changes, in order to transition to a low-carbon future. In particular, the Dingle Peninsula SEC focused on the actions required on the Dingle Peninsula, over the next decade, to meet the targets set out in the Government's Climate Action Plan, which was published in May 2019.

How It Worked

A timeline of the key events;

December 2017—Partnership application submitted to SEAI
April 2018—Dingle Peninsula SEC formally established
4 July 2018—Tender process and approval from SEAI for appointment of consultants to undertake the EMP
4 September 2018—Letter from consultants with their work programme, based on a reduced fee (of €20,000)
14 September 2018—Signed letter-of-offer given to consultants
26 July 2019—Final Report received, after a number of comments had been sent to the consultants based on their draft reports
16 September 2019—Meeting of Steering Committee to consider consultant's EMP. Decision made to use some of the consultants' report (but not all of it) and to utilise further analysis by MAREI.
13 February 2020—Event held to launch the revised EMP.

There were another five Steering Committee meetings held during this time to discuss key developments. Its members included two from Dingle Hub, the SEAI regional mentor, one from ESB Networks, three from Kerry County Council and MaREI was represented by me.

Lesson learnt

The experience of the Dingle Peninsula SEC, along with research from MaREI on other existing EMPs, highlighted a lot of issues with how the processes are structured and quality of the reports [16]. Providing an assessment of the EMP was a useful contribution of the research, but it also created a lot of tension with the consultants. Most SECs will have limited technical knowledge and thus have limited capacity to review the methods used to determine an energy balance baseline. The consultants are used to working on their terms, and were not open to making amendments to the report. This ultimately meant that the SEC decided not to publish their report but instead use the EMP that I drafted. Over the course of the process, my role evolved from documenting what was involved in undertaking an EMP, to advising on best practice in energy planning, to being the service provider.

Providing this expertise was a valuable contribution to the project, which was not limited to the EMP process, throughout the project I made myself available to answer questions on energy and address concerns being raised. This involved meetings with Dingle Hub manager and preparing two pages briefs on issues like smart meters and LED lighting. This service is important as a means to payback partners for their time volunteered to partake in the research. However, it raises questions around legacy and capacity building. Who will offer this service once the research ends? And to what extent is the involvement helping to build capacity and improve energy literacy?

Dingle Peninsula 2030

Background

The establishment of a local partnership or steering group was seen as a matter of best practice both with regard to engaged or transdisciplinary research approaches (Chapter 2) and community energy transition projects (Chapter 4).

The initial vision for the partnership was to create a local decision-making structure that would develop ideas, coordinate projects and build a relationship with key external stakeholders. This was intended to be an open forum, capitalising on the many existing community groups (e.g. Transition Kerry, Transition Dingle, Maharees Conservation Association). However, Dingle Hub were keen to maintain control over what projects were worked on. Instead, a partnership was agreed with four key stakeholders with mutual interest to support each other's work with the Dingle Peninsula community on sustainability projects.

Aims and Objectives

The overall aim of *Dingle Peninsula 2030* was to help create a sustainable future for Corca Dhuibhne (the Dingle Peninsula). The key objective was to work with the local community, schools, business and farming sectors to explore, support and enable the broader societal changes required for the low-carbon transition.

For the research, the goal was to collaborate with local partners to coordinate and design the research. Ensuring that the process was steered from start to finish by the interests and needs of the community under investigation was a research priority. As explored throughout this book (particularly Chapter 6 and 9), this presented some serious methodological challenges as the research had to be flexible and adapt as the project evolved, but was essential in generating impact.

How It Worked

Over 2018, there were several meetings held with Dingle Hub, ESB Networks and NEWKD. MaREI acted as a broker to bring the different organisations together and agree a shared objective. After some discussion, MaREI hosted a workshop in July 2019 to formally agree goals for the collaboration. It was a three-hour discussion broken into six parts.

Session 1: What is Dingle 2030 Trying to Achieve	What is the VISION for Dingle 2030? What are the AIMS and OBJECTIVES?
Session 2: The Rationale for Doing a Case Study	What is the purpose of doing a case study? [to explore, to explain, to describe, to learn, to teach, to advocate, to evaluate, to record/document, to interpret and/or to analyse??] Who will be interested in it?
Session 3: Who is Leading the Case Study/Studies	Do we want to do one case study or a series of separate ones? Will this be a collaborative, engaged piece of research with shared decision-making, shared analysis and shared control over outputs, or is it better to have a more hands-off, observatory overview and analysis? Who will monitor the work?

(continued)

(continued)

Session 4: The Boundaries of the Case Study/Studies	What is the timeline? (November 2017 to November 2021?) What are the research questions we want to answer? What is/are the case study/studies on? Is it just about the 'community engagement' aspects of the Dingle initiatives?—if so, what is in and what is out? What data are we gathering? What methods will we use to record it? Where will we store it? Who will do what?
Session 5: Evaluation and Impact	What is the expected impact of each activity/issue studied? How will we evaluate each activity/issue and its impact?
Session 6: Outputs	What are the case study outputs? Who are they directed at? How will we disseminate them?

Following this, the group met on a monthly basis to provide each other with updates and identify opportunities to work together on different projects, as well as following any new developments in the wider community. These meetings took place in Dingle Hub until March 2020 when they moved online. MaREI coordinated the meetings; setting the agenda, chairing and recording minutes.

Lesson Learnt

We steered the partnership through a process of evaluation and reflection at various stages of the process to capture how it interacted with different networks, what personal and organisational capacities were important to make it work, as well as the benefits and challenges of collaborative governance partnerships for sustainable transitions [17]. Successful collaborations build teams and friendships. They foster trust, creativity, greater insight and understanding. However, collaborations do not happen on their own and they are not always easy. Collaboration requires time and patience to develop an understanding of where partners are coming from.

It is important to build on what is already being done and work with the people who are already active in this field locally.

- Diverse partnerships are key. The combined experiences and expertise from different disciplines (technical and non-technical), with a common mission, stimulate better solutions and problem-solving.
- Include a community development partner in the collaboration to ensure that engagement with the local community is inclusive, participative and empowering and that it helps to contribute to supporting broader societal capacity building.
- It is important, where possible, to document and reflect on the learnings (including benefits, barriers and challenges), in order to inform future practice and planning.

Collaborations need to be set up, nurtured and supported through community-based organisations.

- Time and resources are required to ensure that relationships develop, blockages are worked through and mutual understanding is reached.
- Seed funding for collaborations on sustainability projects, involving multiple stakeholders, should be made more readily available via government departments, agencies and local authorities. Phased grants, where initial funding enables relationship development and proof of concept, are a good model. Allocation of dedicated staff time is required for relationship development and project management (including establishment of governance and communication structures) in agencies, academia, and communities.

References

1. Wellcome Trust, Community engagement – under the microscope (2011). https://wellcome.ac.uk/sites/default/files/wtvm054326_0.pdf

2. C. McGookin, E. Boyle, C. Watson, A. Deane, B.Ó. Gallachóir, Corca Dhuibhne/Dingle Peninsula 2030 learning briefs: reflections on the Engaged Research Approach, (2023). https://www.marei.ie/project/dingle-peninsula-2030/
3. S. Flood, N.A. Cradock-Henry, P. Blackett, P. Edwards, Adaptive and interactive climate futures: systematic review of 'serious games' for engagement and decision-making. Environ. Res. Lett. **13**, 063005 (2018)
4. D. Crookall, Climate change and simulation/gaming: learning for survival. Simul. Gaming **44**, 195–228 (2013)
5. D. Reckien, K. Eisenack, Climate change gaming on board and screen: a review. Simul. Gaming **44**, 253–271 (2013)
6. K. Eisenack, D. Reckien, Climate change and simulation/gaming. Simul. Gaming **44**, 245–252 (2013)
7. A. Krzywoszynska, A. Buckley, H. Birch, M. Watson, P. Chiles, J. Mawyin, H. Holmes, N. Gregson, Co-producing energy futures: impacts of participatory modelling. Build. Res. Inf. **44**, 804–815 (2016). https://doi.org/10.1080/09613218.2016.1211838
8. M. Thomas, T. Partridge, N. Pidgeon, B.H. Harthorn, C. Demski, A. Hasell, Using role play to explore energy perceptions in the United States and United Kingdom. Energy Res. Soc. Sci. **45**, 363–373 (2018). https://doi.org/10.1016/j.erss.2018.06.026
9. S.P. Volken, G. Xexakis, E. Trutnevyte, Perspectives of informed citizen panel on low-carbon electricity portfolios in Switzerland and longer-term evaluation of informational materials. Environ. Sci. Technol. **52**, 11478–11489 (2018)
10. de B.D.M.S.T.B.Ó.G.B.B.E. McGookin C, Corca Dhuibhne/Dingle Peninsula 2030 learning briefs: LED Bulb Swap (2020). https://www.marei.ie/wp-content/uploads/2020/08/Dingle-Climate-LED-Bulb-Swap.pdf
11. 100% Renewable Energy Atlas, Rhein – Hunsrück, Germany (2019). https://www.100-percent.org/rhein-hunsruck-germany/. Accessed 9 Sept 2024
12. B. Ó Caoimh, C. McGookin, Leathinis Corca Dhuibhne – Próifíl Dhéimeagrafach agus Socheacnamaíocha/Dingle Peninsula - Demographic and Socio-Economic Profile, Tralee: North, East and West Kerry Development (2021)

13. Seai, Home/community energy/sustainable energy communities/tools and resources/energy master plan (2022). https://www.seai.ie/community-energy/sustainable-energy-communities/tools-and-resources/energy-master-plan/. Accessed 21 Feb 2023
14. C. McGookin, B.Ó. Gallachóir, E. Byrne, An innovative approach for estimating energy demand and supply to inform local energy transitions, Energy (2021) 120731
15. C. McGookin, D. de Bhailís, B. Tuohy, B.Ó. Gallachóir, E. Byrne, Dingle Peninsula Energy Master Plan (2020). https://dinglepeninsula2030.com/projects/energy/energy-master-plan/
16. C. McGookin, D. de Bhailís, B. Tuohy, B.Ó. Gallachóir, E. Byrne, Corca Dhuibhne/Dingle Peninsula 2030 learning briefs: energy master plan (2020). https://www.marei.ie/wp-content/uploads/2020/11/Dingle-Peninsula-2030_Energy-Master-Plan-Learning-Brief.pdf
17. E. Boyle, C. Watson, C. McGookin, A. Deane, D. de Bhailís, C. McElligott, S. O'Hara, B. Tuohy, B.Ó. Gallachóir, G. Mullally, Reflecting on a collaborative approach to a regional sustainability transition: Dingle Peninsula 2030. Reflective Pract. (2021). https://doi.org/10.1080/14623943.2021.1911794

9

Reflections on the Transdisciplinary Approach

Connor McGookin and Evan Boyle

9.1 What Was the Contribution of the Research Team to the Project?

There were several key contributions that the research made to the project, as outlined in Table 9.1 alongside associated challenges. Most notably is the establishment of a partnership, championing the need for wider community engagement, providing technical and engagement expertise, resourcing a number of initiatives as well as gathering and sharing the learnings from the project through a variety of different outputs.

C. McGookin and E. Boyle, *Engaging Communities in Climate Action*, Energy, Climate and the Environment, https://doi.org/10.1007/978-3-031-93498-8_9

Table 9.1 Summary of research contributions as part of *Dingle Peninsula 2030* [1]

Research contribution	Involvement in the project	Challenges
Establishing the partnership	Coordinating the steering committee	• Getting agreement that a governance structure is needed • Project partners hesitant about opening up the decision-making to a wider group
	Developing good relationships within the group	• Not all partners are equal • Power dynamics
Highlighting importance of community engagement	Providing advice on how to respond to the formation of a local opposition group	Distrust between community and one of the project partners
	Emphasising the need for meaningful community engagement	Lack of funding for engagement activities
	Facilitating community involvement	• Time • Ensuring representativeness • Managing expectations
Expert support	Energy and emissions analysis	Dependence on this support, which ended when the research project ended
	Acting as a sounding board for various aspects of the project	
	Sitting on steering committees	Maintaining scientific integrity and legitimacy
	Presence at public events	Time commitment
	Planning and coordinating engagement activities	
Resources	The research team's time	Being flexible and responding to local needs
		Need to demonstrate impact
	Joint funding applications	University structures not well suited to funding agreements with community partners
	Funding and support for co-hosted events	Finding appropriate funding mechanisms
	Secure funding for the employment of a community engagement/ project manager for the initiative	Lack of core funding for community groups
Capturing and sharing learnings	Reflective learning	Limited time partners have to contribute to co-created outputs

(continued)

Table 9.1 (continued)

Research contribution	Involvement in the project	Challenges
	Wider dissemination, using the learnings to inform policy	Maintaining scientific integrity and legitimacy
Seeing the bigger picture	Highlighting the link between different initiatives	Stakeholders focused on their own goals and projects
	Adding climate change/ sustainability to the local planning agenda	Institutional barriers to collaboration
Evaluation	Helping to build an understanding of reflective learning, evaluation, and the value of impact planning	• Finding right method of evaluation, and ensuring it is utilised • Difficulty determining and attributing impact
	Leading the development of evaluation frameworks, practices and outputs	Balancing research and other outputs

9.2 Key Learnings

Here, we explore five key learnings from the project, focusing on the period from 2018 to 2021. These offer insights into the challenges that emerged and ways in which some of them were navigated, as well as others that remain important areas for future research.

Finding Common Ground

The need for a local coordinating committee was recognised from the beginning as a matter of best practice [3]. It is essential to ensure that all partners have a shared understanding of the aims and objectives of the project [4]. We, as the research team, tried to introduce a democratic decision-making structure with clear community representation at the outset. However, our partners were uneasy about being governed by a steering group too early in the process. Open to their concerns, as an alternative, we settled on a partnership between four key organisations (Chapter 1, Sect. 1.3). This has been a key contribution of the research. Organisations may be focused on delivering their individual projects and meeting their specific goals, while researchers can have a broader view of the bigger picture. The researchers played an essential role in highlighting the links between the different initiatives evolving in the area and acting as a mediator to overcome some initial tensions. We also remained open to local considerations and concerns, taking a flexible approach which was both informed by the academic literature and respectful and reflexive to the community in which we were building the partnership. The benefits of such a collaboration are clear from the breadth of project initiatives (Chapter 5). However, there are crucial gaps in funding for the setting up and coordination of them, which in this case was filled by the research team.

On from how the collaboration is structured, there is the issue of agreeing outputs. In the case of this project, we developed the concept of a 'learning brief' to gather reflections on different elements of the initiative (see Outputs at [5]). Our project partners must contribute to the research on top on their day jobs so time for writing or reviewing is limited. In addition, prior to our involvement with the project, there was a poor understanding of research and what it served to contribute. What has proved successful is when these short documents (learning briefs) are led by the research team but go through a process of review/editing through workshop style discussions in order to ensure they reflect the opinion of the group. Our partners moved to appreciate these as valuable ways to evaluate activities, share learnings, support other community projects, and to inform policy and practice.

Resourcing Engaged Research Projects

We identified a number of funding gaps. Firstly, some of the costs inherent to engaged research are not readily covered under conventional research funding budgets. Existing project funding provides coverage for travel. However, this is intended for attendance at conferences and is insufficient to cover the travelling costs associated with the regular meetings required to manage an engaged research project, or for attending workshops and events in the community. Moreover, there is little available to cover the practical costs of running a workshop or public event. This gap was addressed by our research centre through the development of an Education and Public Engagement mini grant, which offered funding of up to €2,500 for the delivery of specific activities.

There is also the question of how community partners in this project are resourced. At present it is difficult to reimburse people for their time or expenses due to the absence of appropriate mechanisms in the research funding system. Co-production with communities is still relatively new for universities and requires new, more flexible administrative and financial structures as well as timelines. It can be difficult to prepare collaboration agreements that allow funding to go to partners within the community in a timely manner.

The lack of sufficient funding for community development and community engagement slowed the establishment of the partnership, delayed the engaged research and impeded access to, and engagement with, the wider community. A key contribution of the research has been in leading joint funding applications, one of which helped to resource a community engagement role and funding for support from the local community development organisation. This is essential for developing a local governance structure and facilitating wider community participation. However, this only became possible three years into the project. The insufficiency of funding available for such roles makes this a valuable contribution, but there are questions around the role of research as it slips into community development processes.

An additional funding consideration is the role of project management. This is critical for bringing projects from the planning to implementation stage but is poorly resourced. In the *Dingle Peninsula 2030* project, the role was primarily filled by the manager of the remote working space (Dingle Creativity and Innovation Hub) with support from the research team. It means taking on a lot of additional responsibilities alongside normal tasks and is therefore a key constraint for community partners. As noted in the previous paragraph, seeking to address core funding gaps like this is a valuable contribution of the research. However, it also presents a key barrier to engaged research impact, which in this case with regard to climate action, means delivering projects that can significantly reduce greenhouse gas emissions.

Emerging Engaged Research Support Roles

There were two support roles that emerged during the project. Firstly, the role of our engaged research support officer (Clare) in establishing and coordinating the partnership, as well as ensuring good internal communication. The time and effort to maintain effective collaborations is often underappreciated, particularly in community group settings where there is a reliance on volunteerism. Clare offered useful expert advice for our project partners on community engagement, project planning, governance and structure, and how to manage tensions arising from the emergence of a local opposition group. She also provided a 'listening ear' and a sounding board for developing ideas and managing challenges.

Very early in the project, following established best practice [3], our research team identified the need for a committee to coordinate activities in the area. However, reaching consensus on a collaborative partnership was not a simple process. There was initially a tension between our local non-profit partners and the national distribution system operator (ESB Networks) because of their very different financial situations. There was a perception that ESB Networks must have a lot of money available, but in reality, the project budget was very tightly controlled and vast majority of it spent on the technologies used in the trials (home retrofits, solar PV, heat pumps and electric vehicles). Following this bumpy start, it took time before the different organisations built up an understanding and could see the value of working together. A key issue was the fact that the geographical area of interest didn't align since ESB Networks 'Dingle Project' was looking at a region only covering around half the peninsula. Similarly, two central projects of interest for Mol Téic were anaerobic digestion and public transport but ESB Networks was focused on heat pumps and electric cars. This reflects the complexity of the energy system, and highlights the siloed approach to the energy transition that most organisations follow, focusing on their own interests, and neglecting to consider the wider picture.

In addition to this coordination and facilitator role, we think there is a further role to be played in identifying future projects. Having this as a position within the research institute helps to address the fact that current funding structures don't sufficiently resource the initial exploratory stage needed to develop joint funding applications for collaborative projects. Another support role that emerged was a greatly expanded role for the MaREI research centre's Public Engagement Manager (Aoife). Over the course of the project, Aoife became an active team member. She offered regular support with decision-making, coordination of engagement activities, and importantly, led two of the joint funding applications. In addition, the role is important for coordinating the various engaged research projects within the institute, organising training opportunities, and establishing ways for the sharing of learnings among the different engaged research projects. In this case, advancing engaged research is a strategic priority for the research centre, but it raises questions around the expanding role of public engagement personnel more broadly.

Learnings for Engaged Research Practice

There are different stages to an engaged research project. Firstly, an exploratory stage to identify research questions, agree initial goals, methods and outputs. Following this, there is the active period of relationship building, implementation, communication and knowledge sharing. Finally, a key element throughout is reflective learning and evaluation, including post-project impact assessment, which is not easily facilitated in conventional project timeframes. Establishing a good working relationship and research plan, as well as running practical engagements (such as interviews and workshops), requires a significant time investment. A good partnership is built on humility, respect, and common interests.

As a new initiative, developing a shared understanding of the collaboration and how we would work together took time. It was around 18 months before a concrete research plan could be formed following a workshop to agree a vision for the initiative, as well as identifying key research questions and goals. This time investment requires a flexibility in approach that goes against conventional research or project planning that may have strictly timetabled objectives. In addition, we found that it didn't sit well with existing ethics approval procedures in the university (See also Chapter 2). For participatory research approaches it is good practice for researchers to get approval from the Ethics Committee prior to conducting the work. However, this requires a detailed outline of the intended work, which conflicts with the ambition of engaged research to be guided by the partners from outside the university and evolve with the project, in a co-created manner.

The amount of time effective engaged research takes deserves greater recognition, as well as the importance of 'soft' skills in openness, humility, communication, and respect to build up an understanding firstly within the research team and secondly the wider group. Moreover, a point we feel is not often reflected is how the involvement with such a process can support not just an improved professional understanding of the problem, but critically, broadens personal values and world-views. While the time spent travelling to/from meetings or events is a constraint, it is also a useful opportunity for the researchers coming from different disciplinary backgrounds to build a relationship and appreciation for differing perspectives which informs the research direction. In addition, it is a space for both the planning of key topics before meetings and sharing thoughts after.

The issue of time commitment and flexibility is particularly pertinent for early-stage researchers. The engaged research activities are taken on as additional to conventional research activities, as opposed to replacing them, due to the fact that career progression is reliant on conventional academic-based metrics such as journal papers. Stemming from this, there is serious tension between the necessary flexibility in approach and the need to produce academic outputs within the set timeline of a 3- or 4-year PhD. In addition, there is very limited training available on engaged research practices within academic institutes. We participated in a number of facilitation courses from outside the university in the absence of appropriate postgraduate training and learning opportunities. The gap was also in part addressed by the diversity of the core research team, which highlights the importance of a strong interdisciplinary starting point before seeking to reach outside the university. However, despite high-level support for such practice within institutions, current institutional and funding structures present challenges in funding and hiring diverse inter/transdisciplinary teams.

The learning briefs emerged as a useful means to report on different events or outcomes from the project (see Chapter 8, Sect. 8.2.3 for a summary of how they were prepared). These provided an opportunity for critical reflection on areas for improvement, key learnings, and policy recommendations. In some cases, the learning briefs were tied to academic papers, but the majority were not. There is a value placed on the stamp of peer review, but it is often a slow process, and we needed to move quickly to be useful. There is a tension to be navigated between the rigour of the conventional academic process and the desire for wider disseminated outputs, particularly when project collaborators, unfamiliar with the peer-review process, are involved in the writing. Part of dealing with this is using common sense to identify what can be shared quickly (e.g. how different events were run) and what aspects should be withheld until further refined (e.g. policy recommendations based on analysis of the energy system).

Monitoring and Evaluation

The monitoring of progress is seen as very important but equally difficult. This is due to the lack of quality data, different interests of the various partners, societal impacts falling outside conventional metrics that can't be easily determined, and the fact that the research intervention is so short relative to the long-term sustainability transition underway. We suggest that further investigation into how sustainability initiatives diffuse through a community offers an interesting research and evaluation challenge, as does the broader challenge of evaluating and assessing the impact of engaged research.

Progress towards decarbonisation targets needs to be tracked overtime to ensure measures are having the desired impact. Climate change is an ever-evolving super wicked problem given we have limited time to address it, our current actions are making it worse, and not everyone has the opportunity to participate in the solutions. It is thus critical to regularly revisit measures and update plans as required to manage the transition in a timely and equitable manner. In addition, it is important for research to demonstrate impact and evaluate practices in order to draw out learnings for future research.

With regard to tracking progress, there are some difficulties with the energy system models that are used to develop plans. Firstly, the issue of data availability is a significant constraint. This is an issue with both the initial estimate of CO_2 emissions and measuring progress (as discussed in Chapters 3 and 6). In this case, the vast majority of data came from the Census of Ireland, which occurs every 5 years limiting updates to that interval. Secondly, there is the issue of how the models are constructed. In the absence of locally available data for things like building size or details on the car stock, the method to determine energy-related CO_2 emissions relies on a downscaling of national or county/municipal statistics. The use of energy usage indicators like employees or gross value added to determine the starting point subsequently makes it difficult to verify how the area is performing against decarbonisation goals.

From a research perspective, evaluation of projects is generally a funding requirement, but also importantly, can help the development of transdisciplinary research practices by drawing out learnings and failings to understand what works best. A key difficulty is posed by societal impacts that fall outside conventional metrics and cannot be easily determined, as well as the fact that the research intervention is so short relative to the long-term sustainability transition underway. As the framework developed by Pohl et al. shows, there are three distinct stages to a transdisciplinary research project: problem framing, analysis, and exploring impact [6]. There are some signs of progress emerging on the Dingle Peninsula (further discussed in Chapter 10, Sect. 10.2). However, it is only by revisiting it in ten or twenty years' time will we really be able to see the impact that the project had. The standout examples highlighted in Chapter 3 such as Güssing and Samsø are still on their journey to a fossil fuel free future, decades after embarking on the path.

9.3 The Honest Researcher

The ways in which the research team supported the activities emerging on the Dingle Peninsula cover a wide range of different aspects across project planning and coordination, expert input, and a systems thinking approach (Table 9.1). This positioning of the research is at odds with conventional approaches whereby there is clear distinction between the researcher and the researched. We are not independent observers. On the one hand, yes, we are indeed observing how the project evolves and trying to capture what makes it work, but on the other hand we are also working to steer it in a particular direction. A core contribution that research can make is to offer insights on areas of best practice or share learnings from other case studies. How these are presented to the group without taking over the agenda is but one of the balancing acts that the engaged researcher must manage.

There are a variety of different roles fulfilled by the transdisciplinary researcher, as was introduced in Chapter 2 [3, 7, 8]. To make sense of these roles, we pull them together as the traits of the 'honest researcher'. This characterisation of research challenges and builds on Pielke's role for science in policy as *The Honest Broker of Policy Alternatives* [9]. The elements to the honest researcher role may be summarised as follows (Table 9.2).

Table 9.2 The roles of the Honest Researcher

Roles	Description
The Expert	A central component is to offer expertise in a particular field. Within *Dingle Peninsula 2030*, the MaREI research team provided technical analysis on the energy system, support with community engagements and events, monitored the evolution of the project and tracked the impact of the different initiatives and participant experience of the process. The researchers become pro bono consultants: available to address any questions raised by the partners, actively supporting community engagement activities, and preparing a wide variety of outputs

(continued)

Table 9.2 (continued)

Roles	Description
The Communicator	Communicating the decarbonisation challenges and solutions through different mediums, for example: public events, workshop discussions, policy briefs, newspaper articles, etc. Climate and energy are complex/messy problems, so it is essential to be able to communicate key issues in a clear manner
The Listener	Listening to local experiential knowledge and taking it on board. Researchers need to be receptive of conflicting and contrasting opinions in order to build a broader understanding of a problem. This builds on the communicator role above, which would be along the lines of conventional scientific outreach roles seeking to educate or inform the public
The Coordinator	The establishment and coordination of a local governance structure as is necessitated by the engaged research approach. In this case, the role was primarily filled by MaREI's Engaged Research Support Officer and was a significant contribution to the wider project. However, in the long-term it needs to be filled by local actors. Research can help in the formation of such structures as was the case here and in Jühnde (Chapter 3, Sect. 3.4), but cannot be expected to maintain them since research projects typically only last a few years
The Facilitator	Facilitating discussion and debate on the best path forward. The need to open up the energy system modelling and planning process to a more diverse group of stakeholders is well established [10, 11]. Technical experts have an important role to play in these engagements with proper training in facilitation and deliberation. It is important that people have an opportunity to engage with energy/climate experts to pose any questions that they might have and begin a dialogue on the areas of tension as well as the best path forward. More broadly, as has been the case in *Dingle Peninsula 2030*, researchers can play the role of mediator between different stakeholders

(continued)

Table 9.2 (continued)

Roles	Description
The Bridge Builder	Transdisciplinary researchers are uniquely placed when given the opportunity to work both on the ground in local transformation processes and also feed into policy at higher levels. This may help to bridge the communication gap between local communities and national policy objectives, supporting well-informed policy measures. *Dingle Peninsula 2030* has received a lot of national and international attention, which is in a large part due to the research efforts to gather and share learnings. Being a well networked research institution helped due to established relationships with relevant state agencies, public bodies, and industry organisations
The Reflective Practitioner	Contra to Pielke's suggestion that the researcher can be a detached objective observer, the honest researcher works from within the problem. Thus, explicitly recognising and embracing the reality of personal bias. Essential to the credibility of the honest researcher is an openness, humility, and honesty about their personal values and worldview. The researcher's advice is recognised as being framed through their eyes. In addition, to support further development of the field, it is important to critically reflect on the approach taken to identify learnings for future practice, the positioning of transdisciplinary research in higher education and more broadly on the role of research in society

9.4 Disciplinary Perspectives on Transdisplinary Research

We close this chapter with reflections from each of us on the key challenges and benefits of bringing technical and social sciences together to work in a collaborative manner.

9.4.1 An Engineer's Perspective on the Transdisciplinary Approach

What are the Key Challenges Associated with Transdisciplinary Research?

A key challenge faced by transdisciplinary research projects is the ambition of contributing to societal change [12], and the balancing of scientific legitimacy with the process [13]. This need to demonstrate 'success' often means that useful learnings from project failures are not discussed [14]. However, in order to support the development of transdisciplinary approaches it is important to outline failings [15]. One key difficulty faced with the energy planning exercise undertaken was the very flexible and adaptive approach deployed to maximise stakeholder input on the research design process (Chapter 6).

Our local partners lacked the resources to effectively engage with the long-term energy planning process. Without roles dedicated to the *Dingle Peninsula 2030* project, any time spent working on energy and sustainability projects is voluntary and must be taken on in addition to their 'day jobs'. Our partners in the local remote working hub (Mol Téic) were concerned about opening the discussion to the wider community before having a clear understanding of how their role in coordinating a local governance structure would evolve. In light of this, it was deemed important to partner with the community development organisation from the area (NEWKD) to develop a strategic plan for the region informed by an assessment of the demographic and socio-economic challenges, and place energy/climate issues within this. It proved to be very fruitful process and has spawned several important initiatives. However, most of these are not directly related to energy. For example, the most significant outcome from the community meetings held was that Dingle has been chosen as Ireland's representative in the EU 'Smart Rural Areas in the 21st Century' network, which is exploring housing options for a dispersed, ageing population (North East West Kerry Development, 2021). The unstructured nature of the participatory process made it very difficult to translate into the energy system model (as has been previously noted in Chapter 6), but was necessary to ensure local impact and ultimately provides a better understandings of the problem(s).

There is a risk here that the focus on delivering more practical real-world impact will come at the cost of research outputs, which is particularly challenging for early-stage researchers. Given the limited time intervention available to research projects or doctoral students, uncertainty like this that causes delays in workplans and methodological revisions pose a serious concern to the potential for scientific outputs. This emphasises the importance of having an initial stage to co-design the research process and associated outputs [6], which is not generally facilitated by existing funding structures.

A further consideration that emerged through this process is the need for community development rather than engagement. In moving from consultative engagement practices into more collaborative endeavours, researchers slip into community development processes. While this may be of great value to the community of interest or collaboration partners, it raises a number of questions on the role of research and evolving responsibility of researchers. As partners in the project, it is important to build relationships, which requires humility and time to build up trust. However, establishing a good relationship then makes it difficult to manage expectations, as the groups become increasingly reliant on the expert support provided. I regularly offered advice/assistance on energy topics to the local groups, which has been seen as a valuable contribution but begs the question who will fill this role in the future? This makes it crucial to plan for the legacy of the project, and ensure continuation following what is a rather limited intervention by the research project relative to the long-term transition underway. Moreover, a co-production approach rightly seeks to achieve real-world impact and support local needs but when this involves the coordination or co-funding of community development initiatives it begs the question if this is really the role that research is supposed to play within our society?

What are the Key Benefits Associated with Transdisciplinary Research?

There are three elements whereby technical research may benefit from the exposure to other disciplines and a broader range of perspectives: 1) practical support, 2) broader understanding of systems, and 3) critical reflection.

It is important that technical experts seeking to pursue participatory methods don't try and reinvent the wheel, but rather build on existing best practice. Social scientists can provide useful practical knowledge and experience in qualitative data gathering techniques such as surveys, interviews, workshops, etc. They can provide valuable insights into best practice and the importance of carefully planning public engagements. This ranges from providing feedback on surveys, to advise on conducting interviews and support with delivering workshops. Despite a growing demand at a national research policy level for research to be more engaged or action orientated [16], funding for practical expenses like survey companies and facilitators is not available for such research projects. This leaves these duties to be fulfilled by the researchers themselves and thus necessitates collaboration across the disciplines. Likewise, partnering with local actors like community development organisations or environmental groups provides access to existing networks. The community meetings held to discuss the future energy system benefited greatly from the good reputation our local partner had in the area and work they put in to coordinate the workshops (Chapter 6, Sect. 6.3).

Complementary to the practical support with qualitative methods, is the greater depth of theoretical understanding of social, cultural, and political systems. Conventional approaches to energy system modelling generally seek to remove context and complexities in order to develop a quantitative representation of the energy system. However, barriers to the development of renewable energy systems are non-technical challenges that are dynamic and context dependent. This highlights a need for the inclusion of socio-political considerations in the energy system modelling and planning process. There has been much debate in the literature on the ways in which energy system models could seek to integrate different societal elements or why this is fundamentally impossible [17–20]. It seems quite clear to me that a computer tool will never be able to adequately represent the messiness of the real world. The only way an engineer can hope to understand the world within which their solutions are to be built is to go out and engage with the various different actors that inhabit that world.

On a personal level, meaningfully collaborating across the disciplines exposes researchers to differing worldviews and conceptualisations, which fosters a broadening of perspective and makes one more conscious of your personal bias. It facilitates a critical reflection of the engineering discipline and the purpose of design/innovation. This is perhaps the most valuable element, as it forces us to reconsider our place in society, and question if engineering practices are supporting or hindering progress towards ambitions such as the energy system transformation. Furthermore, it fosters a practice of reflection and self-consciousness, which is essential for carrying out engagement activities. Dealing with conflict or contradictions in an interview or workshop setting can be a very delicate matter that requires careful intervention. In such instances, the researcher's role is to act as an objective mediator. However, this position may be easily compromised if the researchers are not well aware of their own inherent biases and values.

9.4.2 A Sociologist's Perspective on the Transdisciplinary Approach

What are the Key Challenges Associated with Transdisciplinary Research?

Despite its increasing relevance, in reality transdisciplinary research has yet to cement a place within academia and remains poorly recognised or understood despite high-level demand from research policy and funding bodies. Consequently, there are tensions inter-alia concerning funding, time constraints, and reporting/ output requirements. Here, I reflect on these findings, building on the discussion above, and offer three key needs to support early-career researchers: greater recognition for outputs other than journal papers, training and developments opportunities, and network building.

Crucially, the obvious importance placed on academic papers and often conflicting metrics of impact must also be acknowledged as an issue [21]. This is of particular relevance for early-career researchers and highlights a need for more flexible evaluation criteria and metrics in the pursuit of transdisciplinary research. As an early-career researcher, the tension related to metrics is one that I must wrestle with as I seek to provide outputs that are at the same time of use to our partners outside of academia, and are considered relevant within an academic context. Our team have sought to achieve this through developing, in parallel, academic papers and learning briefs [22–25]. This, however, has restricted my time with relation to pursuing further upskilling opportunities within the university, which may impact my careers later down the line. Coupled with this, my Department requires a thesis which is delivered as a stand-alone piece of work. Due to this, it was my own responsibility to develop academic papers for peer review and write my thesis. While much of the material was similar in nature it did create a burden with regard to workload. In most cases, the thesis would be turned into papers following the completion of the PhD. This is standard practice within my Department. However, in my case, due to the need to publish public facing learning briefs which were informed by peer-reviewed papers, I had to simultaneously develop these different pieces of work. Timing of outputs is also an issue here. In working with ESB Networks to follow the Ambassador programme within their 'Dingle Project', I provided a qualitative report on Ambassador experiences of integrating new technologies into their homes. Due to the COVID-19 pandemic and associated procurement issues, the quantitative data was not available to supplement my interview findings conducted over a three-year period. The final report not only failed to have this comparative rigour, but also fell outside the timeline of my PhD thesis, meaning a comprehensive report needed to be delivered independent of my doctoral research contributions.

In parallel to these workload issues, as long as TdR fails to contribute towards career development, it will be difficult for capacity and skills to develop in this area, as early-stage researchers must eventually prioritise a narrower focus. Yet, in light of their prominence within EU funding, there is great promise for the future of this research space. Given the high-level mandate in research policy and funding, the responsibility falls to transdisciplinary researchers, who must seize the opportunity to shape an accommodating landscape and overcome existing institutional barriers (See: Chapter 2). An essential element of this will be dedicated training and development opportunities to support those who wish to explore TdR. As demonstrated by the wide range of different terms to describe this work, this is a confusing landscape. This is particularly challenging for early-career researchers trying to understand how their work is placed within the existing literature.

As an early-career researcher conducting transdisciplinary, action-orientated research, I am often faced with a feeling of being an outsider when compared with my disciplinary focused peers. Opening up a wider space for conversions would be fruitful in expanding professional networks. Some of the options identified for strengthening TdR within the institute lie in cross-university gatherings on the topic facilitated through conferences, seminars, publication projects, or university-wide teaching initiatives. These offer important training and development opportunities for early-career researchers, while also giving credence to the need for change by building a network and strengthening transdisciplinary practice. However, in light of the emphasis on conversation as a starting point detached from the institutional pressures inherent within the disciplinary focus of university settings, it is suggested other informal, more socially orientated gatherings are also key, such as postgraduate reading groups or 'meet-ups'.

I have sat between a number of juxtapositions throughout much of this work, struggling throughout to try and bring them closer together. I have been between "going native" and "going academic", perhaps falling at times more towards the former at the cost of the latter. The reciprocal trust-building outlined previously can bring about an over integration in the group. While I found it difficult at times to remain detached from the collaborative groups, I wrote up my findings often in a way that was more distanced. In searching to balance "going native" and "going academic", I need to further develop skills in which I can bring together these two aspects of the process, whereby you are at once part of the group, yet also analysing the group. Finding a voice to represent this was a constant consideration and challenge throughout the project.

What are the Key Benefits Associated with Transdisciplinary Research?

My initial thinking when taking a transdisciplinary approach, working closely with Connor as an Energy Engineer, and non-academic partners, was on further developing epistemological approaches to the pursuit of knowledge, beyond any previous experiences I have had. Sociology, at a theoretical level, can perhaps be criticised for sometimes being a discipline in which the challenge set is who can best describe the dreadful and irredeemable character of late modernity. An awareness of where we are at is warranted. Yet, the solution-orientated approach often used in TdR proves a welcome tonic to many of the issues faced in contemporary society, in this instance with relation to climate change. Alongside this, an increased awareness of my disciplinary blind spots and biases have been highlighted through working alongside engineers, which continually challenges the siloed starting point from which I came. One of the main priorities I had coming into this study was to take an action-orientated approach working closely with a community on the ground. This has been achieved, and many of my most valuable learnings relate to the experiential dimensions and practical considerations.

At a most primary basis, this research is concerned with anthropogenic climate change and the ability of research to play some role in providing solutions to the need for a socio-technical transition to a low-carbon society through collaborative partnership with stakeholders beyond academia. From the position of my own disciplinary starting point, the potential for real-world impact when taking traditional approaches is often limited. However, in undertaking a transdisciplinary, action-research approach the opportunity to play a role in normatively positive processes of change is opened up. As noted by Connor, the solution-orientated and collaborative approach to super wicked problems must be the approach taken across all sectors in response. This moves beyond tradition academic pursuits and pitfalls. In discussing the long-term vision underpinning continued commitment to action research, Brydon-Miller et al. [26], draw on trust-building within community contexts where academic outsiders have often, and often rightfully, been treated with caution. As they note "a key value shared by action researchers, then is the abiding respect for people's knowledge and for their ability to understand and address the issues confronting them and their community" [26] p.14.

This idea of respect for local people and their knowledge has poignancy in the context of the Dingle Peninsula, where anthropologist Scheper-Hughes completed the field research for her 1979 publication *Saints, Scholars, and Schizophrenics*. While the book gained academic success within the fields of psychological and medical anthropology, it was viewed by the Press in Ireland, and the people of the pseudonymised village of 'Ballybran', negatively. Its' publication was seen as an unethical violation of privacy. Some twenty years after the initial scandal, Scheper-Hughes returned to the Dingle Peninsula and the town of 'Ballybran', only to be exiled from the village a few days after her arrival [2]. Although the context of Scheper-Hughes' work, an anthropological investigation into sensitive psychological topics, is far removed from this research project, the lessons learnt give further credence to the approach we have taken. To truly transition, the means through which decision-making occurs must be reanalysed. On a personal level, an awareness of the need for collaborative approaches to decision-making and action to expedite decarbonisation is strongly noted, and in taking a transdisciplinary approach we could act on these considerations in practice.

9.5 Conclusion

The links established within the community and as part of the *Dingle Peninsula 2030* partnership have facilitated the undertaking of the research in a manner that would not have been possible from a detached perspective. Learning from the mistakes of previous researchers such as Scheper-Hughes [2], highlights the importance of building strong interpersonal relationships, founded on trust, good communication, and joint decision-making. Sticking to your word and delivering on promises is essential when undertaking action-orientated research. A clear benefit of the approach is the sense of direct impact and very practical 'usable knowledge' produced.

It is, however, not without its challenges. There is an inherent time pressure for specific outputs within an academic context, which is at odds with the time needed for relationship building. Inversely, often the peer-review process is too slow when compared with the community's need for information. Grey literature, which can bridge this divide, has been developed through our work in the form of learning briefs. This can be recommended as a means through which to disseminate information and findings promptly while academic papers go through the review process. It is not acceptable to focus on one over the other, so again we run into the problem of time. Structuring a collaboration such as *Dingle Peninsula 2030* can be difficult from the perspective of a research centre, funder, and the researchers. However, in light of the widespread impact achieved, the merit of pursuing such approaches is nonetheless clear. Researchers must continue to find creative ways to build collaborations, be open to sharing our failings as well as successes, and learn from each other to support the development of communities of practice around transdisciplinary research.

References

1. C. McGookin, E. Boyle, C. Watson, A. Deane, B.Ó. Gallachóir, *Corca Dhuibhne/Dingle Peninsula 2030 learning briefs: reflections on the engaged research approach* (MaREI Centre, Cork, 2023)
2. N. Scheper-Hughes, Ire in Ireland. Ethnography **1**, 117–140 (2000)
3. D.J. Lang, A. Wiek, M. Bergmann, M. Stauffacher, P. Martens, P. Moll, M. Swilling, C.J. Thomas, Transdisciplinary research in sustainability science: practice, principles, and challenges. Sustain. Sci. **7**, 25–43 (2012)
4. J. Barke, H. Thomas-Hughes, M.E. Howard, Reflections from the field: researchers' experiences of co-production. Res. All **4**, 169–179 (2020)
5. MaREI Centre, Dingle Peninsula 2030. (2022). https://www.marei.ie/project/dingle-peninsula-2030/. Accessed 24 Sept 2024

6. C. Pohl, J.T. Klein, S. Hoffmann, C. Mitchell, D. Fam, Conceptualising transdisciplinary integration as a multidimensional interactive process. Environ Sci Policy **118**, 18–26 (2021)
7. J.M. Wittmayer, N. Schäpke, Action, research and participation: roles of researchers in sustainability transitions. Sustain. Sci. **9**, 483–496 (2014)
8. C. Pohl, S. Rist, A. Zimmermann, P. Fry, G.S. Gurung, F. Schneider, C.I. Speranza, B. Kiteme, S. Boillat, E. Serrano, Researchers' roles in knowledge co-production: experience from sustainability research in Kenya, Switzerland, Bolivia and Nepal. Sci. Public Policy **37**, 267–281 (2010)
9. J.R.A. Pielke, *The Honest Broker: Making Sense of Science in Policy and Politics* (Cambridge University Press, Cambridge, 2007). https://doi.org/10.1017/CBO9780511818110
10. C. McGookin, D. Süsser, G. Xexakis, E. Trutnevyte, W. McDowall, A. Nikas, K. Koasidis, S. Few, P.D. Andersen, C. Demski, Advancing participatory energy systems modelling. Energ. Strat. Rev. **52**, 101319 (2024)
11. D. Süsser, C. McGookin, W. McDowall, F. Lombardi, L. Braunreiter, S. Bouzarovski, Rethink energy system models to support interdisciplinary and inclusive just transition debates, in *Strengthening European Energy Policy*. ed. by A. Crowther, C. Foulds, R. Robison, G. Gladkykh (Springer Nature Switzerland, Cham, 2024), pp.145–157. https://doi.org/10.1007/978-3-031-66481-6_11
12. M. Polk, Transdisciplinary co-production: Designing and testing a transdisciplinary research framework for societal problem solving. Futures **65**, 110–122 (2015). https://doi.org/10.1016/j.futures.2014.11.001
13. J. Köhler, F.W. Geels, F. Kern, J. Markard, E. Onsongo, A. Wieczorek, F. Alkemade, F. Avelino, A. Bergek, F. Boons, An agenda for sustainability transitions research: state of the art and future directions. Environ. Innov. Soc. Trans. **31**, 1–32 (2019)
14. B. Collins, "It's not talked about": the risk of failure in practice in sustainability experiments. Environ. Innov. Soc. Trans. **35**, 77–87 (2020)
15. M.C. Lemos, J.C. Arnott, N.M. Ardoin, K. Baja, A.T. Bednarek, A. Dewulf, C. Fieseler, K.A. Goodrich, K. Jagannathan, N. Klenk, To co-produce or not to co-produce. Nat. Sustain. **1**, 722–724 (2018)
16. E. Boyle, C. McGookin, C. O'Mahony, P. Bolger, E. Byrne, B.Ó. Gallachóir, G. Mullally, Understanding how institutions may support the development of transdisciplinary approaches to sustainability research. Res. All **7**, 1–19 (2023). https://doi.org/10.14324/RFA.07.1.07

17. E. Trutnevyte, L.F. Hirt, N. Bauer, A. Cherp, A. Hawkes, O.Y. Edelenbosch, S. Pedde, D.P. van Vuuren, Societal transformations in models for energy and climate policy: the ambitious next step. One Earth **1**, 423–433 (2019)
18. G. Holtz, F. Alkemade, F. de Haan, J. Köhler, E. Trutnevyte, T. Luthe, J. Halbe, G. Papachristos, E. Chappin, J. Kwakkel, S. Ruutu, Prospects of modelling societal transitions: position paper of an emerging community. Environ. Innov. Soc. Trans. **17**, 41–58 (2015). https://doi.org/10.1016/j.eist.2015.05.006
19. F. Li, E. Trutnevyte, N. Strachan, A review of socio-technical energy transition (STET) models. Technol. Forecast. Soc. Change **100**, 290–305 (2015). https://doi.org/10.1016/j.techfore.2015.07.017
20. W. McDowall, F.W. Geels, Ten challenges for computer models in transitions research: commentary on Holtz et al. Environ. Innov. Soc. Trans. **22**, 41–49 (2017)
21. H. Sauermann, K. Vohland, V. Antoniou, B. Balázs, C. Göbel, K. Karatzas, P. Mooney, J. Perelló, M. Ponti, R. Samson, Citizen science and sustainability transitions. Res. Policy **49**, 103978 (2020)
22. C. McGookin, B.Ó. Gallachóir, E. Byrne, An innovative approach for estimating energy demand and supply to inform local energy transitions. Energy **229**, 120731 (2021)
23. E. Boyle, C. Watson, C. McGookin, A. Deane, D. Bhailís, C. McElligott, S. O'Hara, B. Tuohy, B.Ó. Gallachóir, G. Mullally, Reflecting on a collaborative approach to a regional sustainability transition: Dingle Peninsula 2030. Reflective Pract. **22**, 416–430 (2021)
24. E. Boyle, C. McGookin, C. Watson, A. Deane, B.Ó. Gallachóir, *ESB Networks Dingle Project: Activating Energy Citizenship Through Collaboration, Participation, and Technological Trials* (MaREI Centre, Cork, 2021)
25. C. McGookin, B.Ó. Gallachóir, E. Byrne, Participatory methods in energy system modelling and planning–a review. Renew. Sustain. Energy Rev. **151**, 111504 (2021)
26. M. Brydon-Miller, D. Greenwood, P. Maguire, Why action research? Action Res. **1**(1), 9–28 (2003)

10

Conclusion

Evan Boyle and Connor McGookin

10.1 Fungie and the Flu

Throughout our involvement in Dingle (2018–2022), two major developments impacted the Peninsula. The first of which was of global proportions, the second specifically local. Both, however, play an important role in the story and development of *Dingle Peninsula 2030*. In 2019, Guinness World Records declared a local resident of the Dingle Peninsula the oldest solitary dolphin in the world. Fungie, a male bottlenose dolphin, was first seen in the Dingle harbour in 1983. In the following years, he became a central figure within the Dingle tourism industry with over 10 boats offering trips to see the dolphin every day in the summer. His playful nature, interacting with humans in the water, endeared him to locals and tourists alike. Some anecdotal estimates suggest Fungie brought €2.5 million to the Peninsula each year. In October 2020 he went missing, with marine experts claiming him to have either moved on or to have died. It was suggested at the time that over 50 jobs could

With contributions from—and thanks to—Deirdre De Bhailís

C. McGookin and E. Boyle, *Engaging Communities in Climate Action*, Energy, Climate and the Environment, https://doi.org/10.1007/978-3-031-93498-8_10

be lost following Fungie's departure, coupled with the wider emotional loss felt by locals in the area and others further afield.

Prior to this, in March 2020, the COVID-19 pandemic reached Western Europe and radically altered daily life. The impact of the pandemic had two specific characteristics relevant to our investigation of *Dingle Peninsula 2030*. Tourism numbers plummeted as international and intercountry travel was restricted and as workplaces closed remote working was taken up en masse. Paradoxically, the Dingle Peninsula became at once disconnected and connected to the wider world in a way previously unimagined. Within the context of *Dingle Peninsula 2030,* the vision set out sought to work towards a more environmentally and economically sustainable future for the peninsula. Fungie's death and the impact of COVID-19 are indicative of the need for the area to move towards a more sustainable economy, underpinned by an environmentally driven ethos. *Dingle Peninsula 2030* has set a foundation for this journey. The emergence of remote working as a general practice nationally illuminated Dingle as a viable location for families to live, and with this to build a future.

10.2 What Was the Impact of the Dingle Peninsula 2030 Partnership?

As introduced in Chapter 9 Sect. 9.2, monitoring and evaluating the impact of the project in terms of key decarbonisation targets has proved difficult. The data is scarce, often not available at the granular level needed to understand what's happening in a particular community, and rarely up to date. For example, the Census only comes out every 5 years and national energy and emissions are reported 2 years after the fact (i.e. 2022 figures are released in 2024). Even if all these issues could be resolved, there remains the question of attribution, and to what extend we can clearly link research activities with impacts. Nonetheless, there are some success stories worth highlighting here.

One of the biggest impacts stemming from the project has been the sustainable mobility initiative. This has seen the introduction of two new Local Link bus routes, providing a vital connection for some of the communities, and with the introduction of electric buses in 2023, reducing transport emissions. With the new electric buses carrying ~ 3,000 passengers per month, the emissions savings are estimated to be ~ 80 tCO_2, equivalent to taking 40 cars off the road.[1] A small dent in the 26,840 tonnes CO_2 emitted from private travel on the Dingle Peninsula in 2018 (Chapter 6 Sect. 6.5), but a significant step in the right direction.

Looking at the uptake of new energy technologies there are positive signs. Since 2021, households on the Dingle Peninsula have installed 151 solar PV systems, 55 heat pumps, 71 EV chargers, and 480 homes underwent energy efficiency improvements such as attic or roof insulation.[2] This growing interest in home energy upgrades has brought economic opportunities, local start-ups such as Solar Beo and DC6 Technologies were boosted by contracts during the ESB Networks Dingle Project [1]. It is estimated that there are now 192 jobs in local sustainability-orientated businesses [2].

Another major development has been the emergence of two new Sustainable Energy Community groups on the peninsula, one for the Tourism & Hospitality sector and another for Farming [2]. The Tourism and Hospitality SEC have completed 20 solar PV installations, and 18 within the Dairy Farmers SEC[2]. There have also been several assessments on the feasibility of anaerobic digestion, and a lot of interest in the shortening of supply chains by giving hospitality direct access to local produce. While the case for both is clear, and together it presents a nice synergy with waste from food fed into the digestor, these efforts have been stalled by Ireland's tightly regulated meat processing chain, and a lack of capital funding for larger energy plants.

In 2023, the Dingle Peninsula was chosen to be Kerry County Council's Decarbonisation Zone. A decarbonisation zone is a region where different climate mitigation, adaptation, and biodiversity measures are to be showcased. The selection of the Dingle Peninsula as Co.

[1] Estimate prepared for Dingle Hub as part of the preparation of a new Decarbonisation Zone Strategy for the Dingle Peninsula in 2023. Unfortunately, it was not possible to verify this estimate with data on actual passenger numbers and fuel demand.

[2] Based on email correspondence with Dingle Hub, Sustainable Energy Authority of Ireland data is not yet available but is due to be published in early 2025.

Kerry's Decarbonisation Zone recognises the strong foundation built there, and the potential the region has to demonstrate leadership in the transformation of our energy system away from fossil fuels.

While it is important to celebrate the success stories, and impact seen to date, there is still much work to be done. We are remained of the importance of emissions accounting and tracking progress. The contribution each of the successes outlined here will make to the overall emissions reduction objective is minimal but the sum of the parts becomes increasing more impactful as momentum grows. Ongoing evaluation can provide good news stories that will encourage others to get involved, while also importantly highlighting areas for improvement and identifying which sectors need to be targeted for mitigation measures.

10.3 Merging Top-Down and Bottom-Up Approaches to Sustainability

Internationally, democracy is at a point of crisis, particularly with relation to distrust in political institutions [3, 4]. The wide range of existential threats posed to twenty-first-century society, such as artificial intelligence, climate breakdown, the biodiversity crisis, and geopolitical tensions, are positioned within the context of the socio-political crisis.

Re-imagining the institutions which structure society is a pertinent project [5]. As E.O. Wilson notes "*the fundamental problem of humanity is we have Palaeolithic emotions, medieval institutions, and accelerating God-like technology*" ([6], p.34). While the former is beyond influence, the latter two must be aligned moving forward. The political, economic, and social structures which have become embedded in western society are no longer fit for purpose. New macroeconomic systems which can account for the multitude of issues faced are being proposed [7, 8]. Politically, experimentation with new forms of governance has been seen with the rise of citizens' assemblies globally as a form of deliberative democracy [9]. In relation to climate change and issues of environmental sustainability, the Irish citizens' assembly on Climate Change in 2016 represented an important milestone in deliberative democracy for climate action [10, 11]. Anthropologically, we can look to Clastres' [12] depiction of power relations in the native rainforest cultures of South America, and its decentralised approach, away from its coercive potentialities. In ancient Grecian culture, as noted by many in relation to the contemporary citizens' assembly model [13, 14] participatory democratic processes were central to political life.

These political ideas, at a macro level, can be related to more specific micro-processes. Within this context, the merging of top-down and bottom-up approaches to sustainability can operate as niche experiments for imagining and developing new political structures. This research project, while in no sense a panacea to the ills of contemporary democracy, has offered forth some findings and inspirations for participatory collaborations across different groups, in a transdisciplinary approach. Such an approach presents insights with relation to governance for sustainable transitions and deliberative democracy more broadly. There is a fundamental dualism within notions of participation, whereby one must be externally empowered and internally able to participate in social and political life (Horvath & Szakolczai, 2018), *Dingle Peninsula 2030* can act as an example where participation and multi-actor processes assist in building capacity to enact these two requirements. In the first case through building collaborative networks and partnerships and secondly through developing personal capacities.

10.4 Final Words from Dingle Hub

In concluding the writing of this book we organised a final meeting with Deirdre, the manager of Dingle Hub. It was held online in October 2024 and was an opportunity to reflect upon our collaboration and to look forward to 2030. Dingle Hub was the key driver for *Dingle Peninsula 2030* and continues to advance rural sustainability on the Dingle Peninsula. Deirdre is central to this, acting—as it has been described to her—as a community designer. The last words on the project are rightfully hers. We have divided this section over three key sub-sections, with some quotations provided from our conversation with Deirdre.

10.4.1 Legacy of Dingle Peninsula 2030 and the Continuation of the Work Through Dingle Hub

The success of Dingle Peninsula 2030 was borne out of the collaboration of the four main projects partners, led by Dingle Hub. Crucially it was supported, championed and developed by local people on the peninsula with motivations to deliver different initiatives in their region to enable more sustainable practices to be established across a range of sectors. As the research partner, MaREI played a key role in providing the evidence base to support this work and was able to help establish a foundation from which activities on the peninsula could grow and flourish through the diffusion of sustainability. Deirdre, in reflecting upon this, highlighted the key role played in documenting the work which was undertaken:

> *Reflecting on the Dingle Peninsula 2030 partnership, we have a good track record for delivery, but it was the team from MaREI that documented the work and gave us credibility through the library of learning briefs and the peer-review journal articles. It shows that there was a strategy to it which is well documented and can be shared with other areas. Building that credibility has led on to the strong relationships with external stakeholders that we've been in and that has led on to funding successes as well.*

The base established through Dingle Peninsula 2030 can be useful referred to as first-generation projects. Since our direct involvement, MaREI has continued to partner with Dingle Hub on projects such as, for example, ENPOWER, including new research partner EPRI (Energy Power and Research Institute). This European research consortium aims to design, develop and demonstrate social sciences and humanities driven methodologies, and interactive, closed-loop ICT tools and services for energy activated citizens and data-driven energy-secure communities towards a consumer-centric energy system. The Dingle Peninsula acts as the Irish pilot site among five others across Europe, focusing on three activities spanning the tourism, agriculture, and transport sectors to address technical and social aspects. Dingle Hub is also continuing with projects related to public transport, local short supply chains and anaerobic digestion, fitting sustainability initiatives to local needs and opportunities in line with the scaling deep model which they have designed.

10.4.2 The Need for Sub-County Structures to Facilitate Multi-Layered Projects

A continuous challenge which is faced by Dingle Hub is the funding landscape. Currently, policies and strategies are set nationally, but the burden of implementation is left to the local level where capacity is severely restricted in relation to delivery. Dingle Hub, in 2024, was funded through a mixture of providing commercial services and philanthropic support. The need for sub-county structures to help support the emergence of multi-layered projects such as that represented by Dingle Hub is a continuous point of consideration for Deirdre in her role as manager:

> *There should be exchequer support in the mix as well. It's not right that there isn't. This year, we are entirely funded by commercial services and philanthropy. That does not feel morally right or right from a policy perspective for the country. The model could be (roughly) a third exchequer, a third commercial, and a third other sources or projects. If there was 30% support from an exchequer fund it becomes a workable model. But that's only once you reach scale. We can do those commercial services now because of the track record we've built. You still need a mechanism for communities starting out in this space, where they could have a level of independence and to build their own income streams for delivering services.*

In 2024, Dingle Hub made a submission to the public consultation on Ireland's draft Climate Action Plan. The submission focused singularly on the need to enable local funding of community initiatives. Current policy assumes the ability of volunteer capacity and capability to apply for, manage, administer and monitor different funding schemes offered. Coupled with this, they must undertake the challenge of continuous engagement with the wider community. As noted in the submission, one of the consistent failings in all State supports to date is a failure to acknowledge (and resource) some sustained form of local community group or company limited by guarantee that can engage with the citizens in the community and coordinate projects. Due to the scale of the challenge presented both in relation to climate change and biodiversity, the submission suggests that a more streamlined approach should be developed by Government for supporting communities to act on these twin challenges. Such an approach should in the first instance be aligned and focused on the needs of local communities as opposed to administrative architectures as designed by governmental departments and agencies.

10.4.3 Looking Ahead to 2030

The Dingle Peninsula is a collection of communities of interest with established boundaries and articles of differentiation. It is an area conflicted by historical and global forces bringing about the proliferation of unsustainable practices across numerous sectors. Yet, among all this disparity and degradation is the amazing natural beauty and the modes of unification which bring people together; the Connor Pass and the Slea Head Drive, the rugged landscape and the inclement weather. Continuing from the work of Brent (2004) community can be understood as desire. Dingle Peninsula 2030 is a desire for community and sustainability. The Dingle Peninsula, like much of the world, is in a moment of transition. While there is much cause for concern, the original foundations of Dingle Peninsula 2030 and the continued work of Dingle Hub and the wider region may assist in evoking this desire for community across the region in the years ahead. In looking forward, Deirdre provided some final reflections on her hopes, firstly in relation to specific project areas and then more broadly;

> *Firstly, there's huge interest in the sustainable mobility work. In the next 5 years, I would like to think that we will have achieved an impactful level of reduction in our private car use. At least to have that network of quiet lanes in place, that walking and cycling has become safer. On the local food, an absolute dream is that we can unlock that and in a five-year time frame have viable operational structures established here and access to local food.*

The impact of that is more local farms being financially sustainable into the future. And on the anaerobic digestion, I would love to see that we have our demonstration project up and running by 2030 and are in a position to have the commercial model, the business case that's suitable for investments for a commercial-scale plant here which will unlock decarbonisation for us.

Overall, I have described myself as ridiculously optimistic. But I am, I can see the amount of people who want to get involved on the ground. Of course, there are lots of people who haven't engaged at all and say "this just doesn't impact me in any way". But I see so many people joining. The different activities, the different action. Taking action themselves, wanting to do it. And that makes me very hopeful for the future.

References

1. C. Watson, E. Boyle, C. McGookin, A. Deane, B.Ó. Gallachóir, D. de Bhailís, B. Tuohy, S. O'Hara, C. McElligott, (2022) Corca Dhuibhne/ Dingle Peninsula 2030 learning briefs: lessons learnt and recommendations for policy and practice. https://dinglepeninsula2030.com/wp-content/uploads/2022/06/Lessons-Learnt-and-Recommendations-for-Policy-and-Practice_Final.pdf
2. Deane A. Corca Dhuibhne 2030/Dingle Peninsula 2030 impact case study. 2023. https://www.marei.ie/wp-content/uploads/2023/09/Dingle-Case-Study-2023.pdf
3. V. Hart, *Distrust and Democracy: Political Distrust in Britain and America* (Cambridge University Press, Cambridge, 1978)
4. A. Applebaum, *Twilight of Democracy: The Failure of Politics and the Parting of Friends* (Penguin UK, London, 2020)
5. C. Harris, I. Hughes, Reimagining democracy in an era of deep transition. Ir. Stud. Int. Aff. **31**, 71–89 (2020)
6. L. Else, Talking to EO Wilson: ants, apps and altruism. New Sci. **214**, 34–35 (2012)
7. A. Pettifor, *The Case for the Green New Deal* (Verso, London, 2020)
8. K. Raworth, *Doughnut Economics: Seven Ways to Think Like a 21st-Century Economist* (Chelsea Green Publishing, London, 2017)

9. C. Chwalisz, A new wave of deliberative democracy. Carnegie Eur. **26**, 1–6 (2019)
10. T. Forsberg, Citizens' assemblies: a potential transformative method for addressing the wicked problem of climate change: a case study of the 2016 Irish Citizens' Assembly, (2020). https://www.diva-portal.org/smash/record.jsf?pid=diva2%3A1455021&dswid=8539
11. L. Devaney, P. Brereton, D. Torney, M. Coleman, C. Boussalis, T.G. Coan, Environmental literacy and deliberative democracy: a content analysis of written submissions to the Irish Citizens' Assembly on climate change. Clim. Change **162**, 1965–1984 (2020)
12. P. Clastres, R. Hurley, A. Stein, *Society Against the State* (Blackwell, Oxford, 1977)
13. J.S. Fishkin, Deliberative democracy in context: Reflections on theory and practice. In: *Deliberative Mini-publics: Involving Citizens in the Democratic Process* (2014), pp. 27–39
14. M.E. Warren, H. Pearse, *Designing Deliberative Democracy: The British Columbia Citizens' Assembly* (Cambridge University Press, Cambridge, 2008)

Index

C. McGookin and E. Boyle, *Engaging Communities in Climate Action*, Energy, Climate and the Environment, https://doi.org/10.1007/978-3-031-93498-8

The manufacturer's authorised representative in the EU is Springer Nature Customer Service Centre GmbH, Europaplatz 3, 69115 Heidelberg, Germany. If you have any concerns regarding our products, please contact ProductSafety@springernature.com

Printed and bound by CPI Group (UK) Ltd, Croydon, CR0 4YY
01/10/2026
02225954-0004